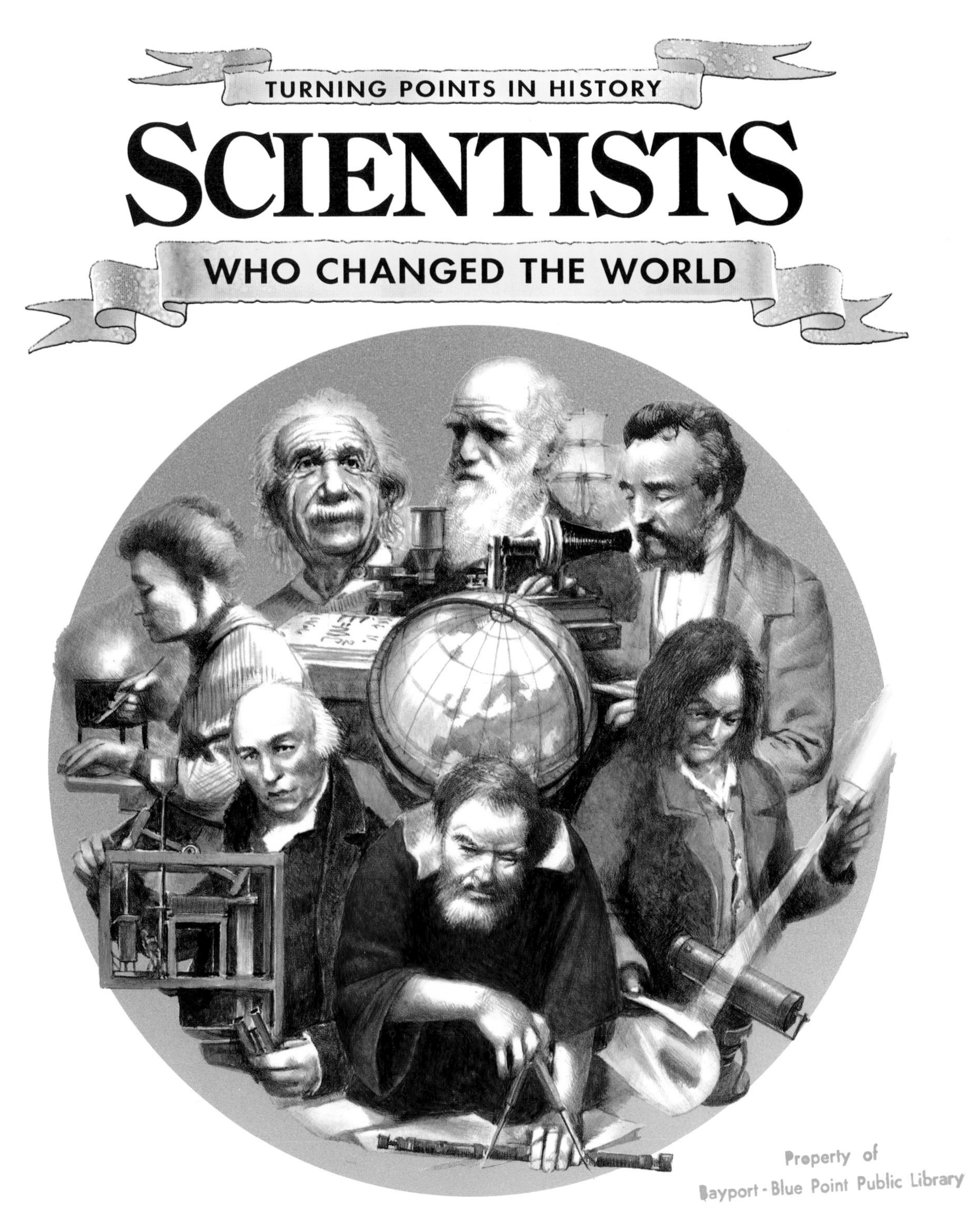

Philip Wilkinson & Michael Pollard
Illustrations by Robert Ingpen

CHELSEA HOUSE PUBLISHERS
New York • Philadelphia

First published in the United States by
Chelsea House Publishers, 1994

First Printing
1 3 5 7 9 8 6 4 2

Simplified text and captions by **Michael Pollard**
based on the *Encyclopedia of World Events*
by Robert Ingpen & Philip Wilkinson

Editor	Diana Briscoe
Project Editor	Paul Bennett
Designer	Design 23
Design Assistant	Victoria Furbisher
DTP Manager	Keith Bambury
Editorial Director	Pippa Rubinstein

ISBN 0–7910–2763–5

Printed in Italy.

Contents

The Invention of Gunpowder 10
Kaifung, China, A.D. 1280
Printing with Movable Type 15
Johannes Gutenberg, Mainz, Germany, 1453
Galileo & His Telescope 20
Galileo Galilei, Padua, Italy, 1610
Newton & the Laws of Gravity 24
Isaac Newton, London, England, 1687
Watt & the Steam Engine 27
James Watt, Glasgow, Scotland, 1769
Preserving Food 32
Donkin & Hall, London, England, 1813
The Invention of Photography 35
Louis Daguerre, Paris, France, 1839
Darwin's Theory of Evolution 39
Charles Darwin, Galapagos Islands, 1859
Lister the Germ-killer 44
Joseph Lister, Edinburgh, Scotland, 1867
Bell & the Telephone 48
Alexander Graham Bell, Boston, Massachusetts, 1877
Marie Curie & Radium 52
Marie Curie, Paris, France, 1903
The First Powered Flight 55
The Wright Brothers, Kitty Hawk, N. Carolina, 1903
Henry Ford & His Model T 60
Henry Ford, Detroit, Michigan, 1913
Einstein's Theory of Relativity 65
Albert Einstein, Berne, Switzerland, 1915
The Inventors of Television 69
John Logie Baird, London, England, 1925
Building the First Computers 72
Alan Turing, Bletchley, Bucks, England, 1943
The Atom Bomb 75
Manhattan Project, Los Alamos, New Mexico, 1945
The Decoders of DNA 79
Crick & Watson, Cambridge &
Wilkins & Franklin, London, England, 1953
The Launch of Sputnik I 82
Tyunatam, Russia, 1957
The First Men on the Moon 86
Aldrin, Armstrong & Collins,
Cape Kennedy, Florida, 1969
Further Reading 91
Index 91

Introduction

Today, the products of the great scientists and inventors are obvious everywhere – from video games to digital watches, cars to computers. But as long as humans have lived on the Earth, there have been people who have used their ingenuity to try to improve their lives. These unsung geniuses are lost in the mists of time. We take inventions like writing, tools, and the plow for granted now. But at the time they appeared, they brought about enormous changes.

Farming with a plow produced a situation in which people could grow more food than they needed. This freed other people to do different jobs, like making pots or items from wood, and trade could begin among the various groups. The craftworkers and traders could live together in towns, creating together an entirely new, city-based culture, which became what we know as civilization. Trade in turn encourages transport, leading to better boats, wheeled vehicles, and the stirrup, vital for horse riding.

One of the greatest inventions of all was movable type for printing, which was pioneered by Johannes Gutenberg about 1450. The printing press made it possible to produce books in large numbers with great ease and at high speed. Gutenberg's invention caught on very quickly, with the result that information could travel easily around the world. This helped other scientists – new ideas could be publicized, and one scientist could benefit from the discoveries of another.

The speed of technological change also greatly increased. Suddenly, inventions seemed to come more quickly. Some, like the steam engine, brought great changes in the lives of ordinary people. As a result of the work of many engineers and scientists, it led to developments in mining and manufacturing that produced the industrial revolution, as well as making the railways possible.

The railways themselves were just one in a series of breakthroughs that

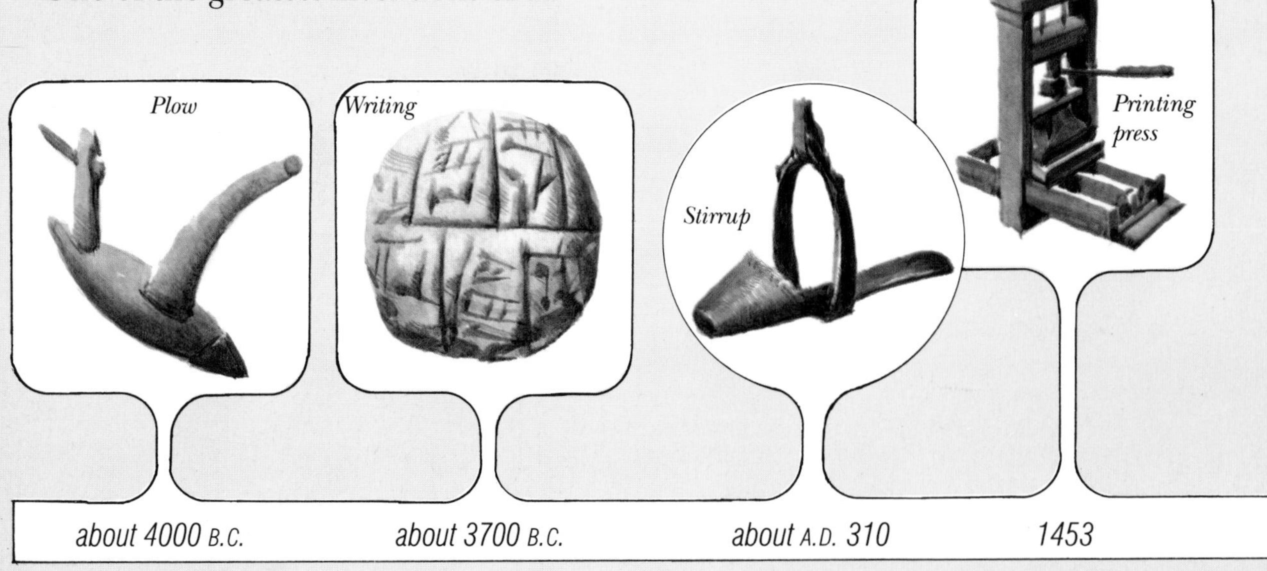

have vastly improved our ability to get around the world. They were followed by the pioneering work of Daimler and Benz on motor vehicles, of the Wright brothers on the airplane, and of armies of Russian and American scientists on spacecraft.

There have also been great advances in sending messages and signals of all sorts around the world, from Bell's telephone to the television. Five hundred years ago few people knew what was happening a hundred miles away. Today we are kept constantly up-to-date on news from all over the world, beamed from satellites. It seems as though the world has shrunk so that we can see it all at a glance.

Indeed, space travel has allowed us to see the Earth as a whole for the first time. This view of our planet has become a symbol of hope that people will use science to look after the Earth for its future inhabitants.

Philip Wilkinson

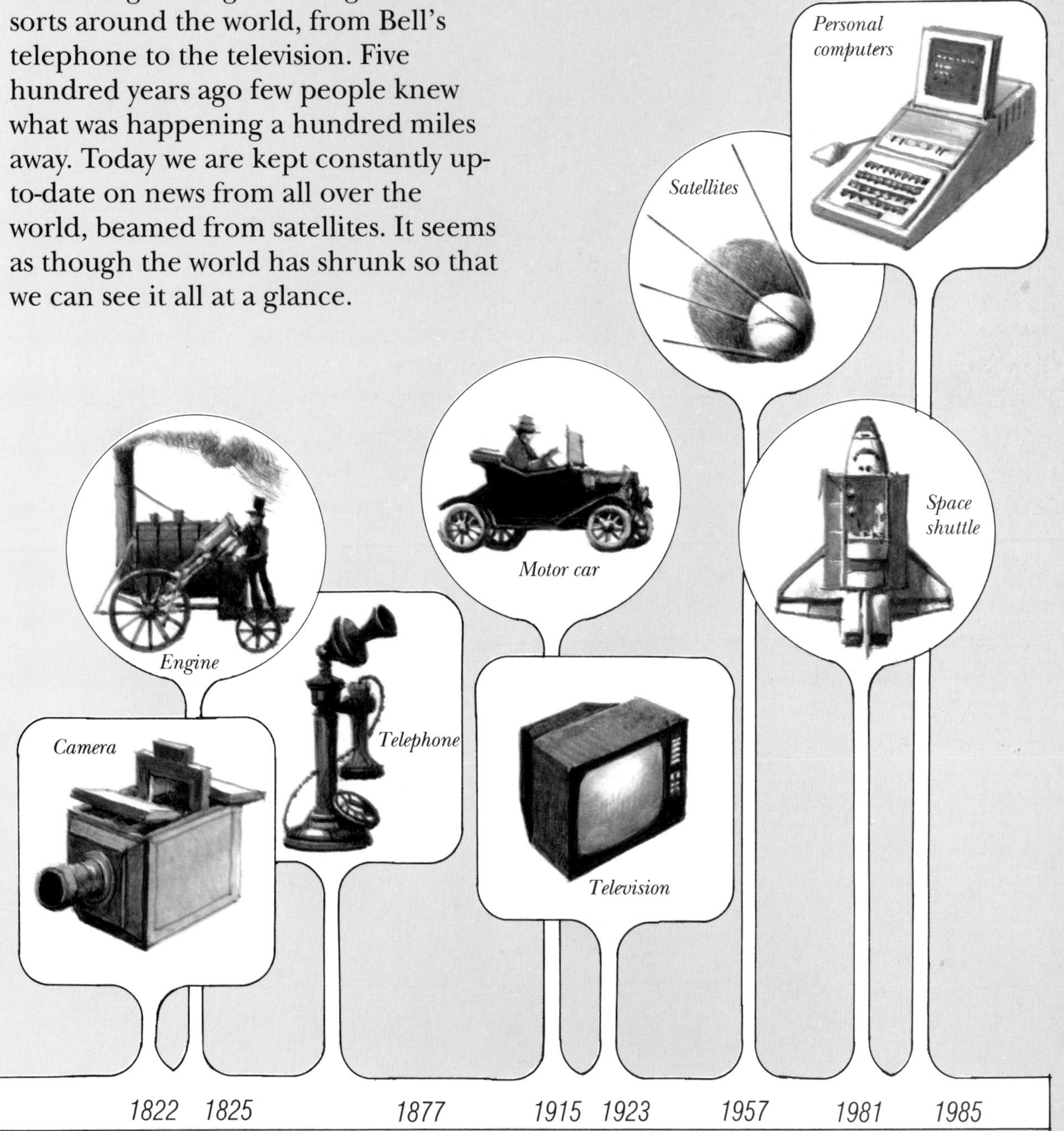

The Invention of Gunpowder

Crash! Thud! Crash! Thud! As terrified townspeople cowered behind their city's wall, they must have thought their attackers could command the thunder and lightning. Those first cannons, powered by gunpowder, changed the way that wars were fought forever – the weapons had become more important than the soldiers.

1200 1300 1400 1500 1600 1700 1800 1900 2000

1280 Kaifeng, China

No one knows where or when it happened, but there was a day – about 700 years ago – when the people of a walled city were sheltering safely in their homes, confident that the enemy at their gates could do nothing to break in. Their city walls were strong enough to resist any battering ram. Soldiers on watch on the ramparts would prevent enemy troops from climbing them. Their city was as safe as any city could be.

Terror from outside

Suddenly, the calm was broken by a deep roaring sound beyond the walls. Seconds later, there was a crash as something heavy smashed into the stonework and sent cracks running along the joints. Another roar, another crash, and the walls began to crumble.

There were clouds of dust and smoke, and the sound of men shouting. Out of those clouds came the invading army, its way into the city blasted wide open.

Force of destruction

Those city walls had been breached by shot fired from cannons in one of the first attacks of its kind in history. No battering ram or trebuchet – a kind of huge catapult – had the destructive force of the gunpowder that was packed into the barrel of a cannon.

No one knows exactly when gunpowder was invented or when it was first used to fire cannons. The first people to make gunpowder, probably around the year A.D. 1000, were the Chinese. They discovered that certain substances, such as sulphur, saltpeter, charcoal and pitch, produce a fast-burning mixture which explodes with great violence if lit in a confined space. The original inventors of gunpowder may have

Because each cannon was cast as a one-off, no two cannons were alike. The master gunner had to know the quirks of each of his guns to use them to best advantage.

intended it for use in fireworks, but by 1288 the Chinese were using cannons in their battles against invaders from the north.

From then on, the use of gunpowder seems to have spread amazingly fast, perhaps through Russia, to Europe. It was not simply a matter of passing on the details of the explosive mixture. The technique of using it to fire shot – either iron or stone – by ramming it into the end of a barrel, loading the shot and then setting light to the charge also travelled like fire across the world. With it, the nature of war changed for ever.

The cannon's roar

In 1338 England and France began what has become known as the Hundred Years' War, which lasted until 1453. One of the first times cannons were used in Europe was by the English army during the Battle of Crécy in 1346. In the following year they were used to bombard the town of Calais into surrender.

Probably the main value of the first cannons was to terrify the enemy. But as they became more accurate, and the gunpowder mixture more reliable, cannons became deadly weapons in their own right. By 1453 the Turks were able to reduce the mighty walls of Constantinople to rubble after a relentless eight-week bombardment.

The first handguns

By this time, gunpowder was also being used in handguns. The first of these were rather like miniature cannons, which were carried over the shoulder. They were fired in the same way, by ramming a gunpowder charge down the barrel, but the shot they fired was

The Chinese used gunpowder in fireworks from about 1025.

FASCINATING FACTS

Roger Bacon, an English friar, was the first European to mention gunpowder in his book Opus Majus, *written about 1268.*

The oldest known gun was made in China about 1288. The oldest known European cannon was found at Loshult in Sweden. It was made in about 1350.

During the Turkish bombardment of Constantinople in 1453, nearly 4,000 stone shots were fired at the city walls. Each shot weighed over 800 pounds (400 kg) and was fired by 300 pounds (150 kg) of gunpowder.

A Swiss businessman, Henri Dunant, founded the International Red Cross movement after seeing the results of gunfire at the Battle of Solferino, between armies of France and Austria in 1859.

Big Bertha was the name given to a German gun which bombarded Paris in World War I. It fired explosive shells weighing 363 pounds (115 kg) over a distance of 75 miles (122 km). Big Bertha caused 256 deaths.

made of lead. They were not very effective, because they could only be used at a range of up to about 30 feet (ten meters) and were slow to load and fire. It was the awesome destructive power of the cannons that changed history.

Gunners for hire

On land, people could no longer feel safe behind the ramparts of their castles or in walled cities. A knight or warlord could no longer rely on the size of his army to protect him and fight his battles, and his long training in the tactics of battle rapidly became out-of-date.

Handling cannons was a job for experts, and a master gunner with an efficient team of about ten men would hire themselves out to the prince or general paying the highest price.

From siege to battlefield

The battles of the Middle Ages were often sieges, like those of Calais and Constantinople. An enemy army surrounded a city and starved it into submission. However, once the cities acquired cannons as well, sieges often became stalemates. The besiegers' cannons could damage city's walls, but the city's cannons could prevent the attackers from storming the breach.

After about 1500 and for the next 400 years, land battles were more often

A walled city is being bombarded by cannons during a siege. The advantage of cannons over battering rams and other earlier "siege engines" was that they could be set up at a distance, out of range of the enemy's spears or arrows. There was little the defenders on the battlements can do except to take cover and wait for the assault, which is just starting across the bridge.

"set pieces" in which the opposing sides would form lines of heavy guns in front of their foot soldiers. The battle would begin with a heavy pounding by both sides' guns.

The guns of empire

The effect of the invention of gunpowder on sea warfare was just as dramatic. By 1500 the major seafaring nations were arming their ships with cannons. Henry VIII of England's flagship, the *Henri Grâce à Dieu,* which was built about 1515, was armed with 186 guns. But it was too heavy to be much use in battle. The more usual number for a battleship was about 100 guns.

It was this enormous firepower that enabled the British, Dutch, French, Portuguese and Spanish fleets to establish their empires. The native peoples of Africa and the Americas had no defense against shot, whether it came from ships bombarding them off the coast or from soldiers armed with handguns. It was an unequal contest which the European empire-builders could not fail to win.

Technology rules

So from about 1400 onwards power in the world was no longer exercised by leaders who could gather round them a force of the most fearless and skillful individual soldiers. Power was in the hands of generals and admirals of specific nations who could command the firepower of gunpowder.

War became more deadly, while the risks of death or injury became much greater. And victory almost always went to the side with the better technology.

Longbows and crossbows were the long-range weapons of the Middle Ages. Longbows were more effective, but needed much more training and practice by their users.

Printing with Movable Type

There are more than 774,000 words in the Bible, and a monk writing with a quill pen could take from five to 30 years to make one copy. But Gutenberg printed 300 copies in less than one year, using his new invention. There was an information explosion as printing presses spread – over 35,000 different books were published within 50 years!

1200 1300 1400 1500 1600 1700 1800 1900 2000

1453 Mainz, Germany

Printing with movable type was one of the many inventions which began with the Chinese. But because China was not in touch with the Western world, news of it did not spread beyond the Far East. It was "re-invented" again in Europe about 400 years later, and this time spread throughout the world.

Patient scribes

Until books could be printed, the only way of producing them was to copy them out laboriously by hand with a quill pen. This was the work of scribes, who were usually monks. It could take years to copy a book, which made books so expensive that only the very rich could afford them. As there were so few books made, new ideas and knowledge spread very slowly. Most people never saw a book, except in church, in the whole of their lives.

Some printers tried carving the words of books, together with illustrations, on to wooden blocks and printing from these. But this was almost as slow as copying, and the wooden blocks soon wore out. What was needed was a method of making separate pieces of metal type for each capital and small letter, each numeral and punctuation mark, and then assembling them to make the words on a page of a book. The type could then be inked and pressed against a sheet of paper in a press. This machine was adapted originally from the kind of press used to crush fruit and extract its juice or oil.

est littera psalterij

est littera psalterij

Johannes Gutenberg (c. 1398–1468) was born in Mainz, Germany. He struggled for many years to perfect his method of printing.

Who was first?

No one can be sure who solved the problem first. In Germany, the Netherlands, Italy and France there were craftsmen all aiming for the same goal. It often happens with new inventions that there is a

race between a number of people all working along the same lines. Most experts today agree that Johannes Gutenberg (GOO-ten-burg) was the winner of this particular race.

Making words out of metal

Gutenberg was born in Mainz in southern Germany and learned his metalcraft at its mint. Coins were produced by heating gold and silver and pouring the metal into molds. It was probably from this process that Gutenberg got the idea of casting individual pieces of type out of metal.

In 1428, Gutenberg moved to Strasbourg and set up as a goldsmith. He was probably already working in secret on his movable type project. But he needed money to buy metals and equipment, and by 1448 he was back in Mainz borrowing money from a businessman called Johann Fust.

First time in print

At last, about 1450, Gutenberg began printing his first book. It was a 641-page Bible, in Latin, and he printed about 300 copies. Sadly, Gutenberg's triumph was snatched away from him. He failed to pay back his debts to Johann Fust, and Fust took over his printing business. This continued and prospered, but Gutenberg retired and little more is known about him.

While Gutenberg checks a page of type (left), his craftsmen continue their work. The compositor (center) is selecting letters from the wooden type-case to set them in a composing stick. While another worker operates the press, the man in the background is checking newly printed sheets for mistakes. Above him, printed sheets hang up to dry.

FASCINATING FACTS

The first person to use movable type was a Chinese inventor, Pi Sheng, in the 1040s.

The oldest surviving European example of printing from movable type is a scrap of paper with eleven printed lines on it. It probably dates from 1442.

Between 1453 and 1500, European printers produced over 35,000 different books.

The earliest known advertisement for printed books was issued by Heinrich Eggestein at Strasbourg in 1466.

In 1516, an Italian printer at Genoa published a Psaltery *(Book of Psalms) printed in Hebrew, Greek, Arabic and Syriac. Each language needed a different alphabet.*

The base of a printing press is called the bed. When newspaper printers are ready to start printing they say they are "putting the paper to bed."

The first printing press in the Americas was run by priests in Mexico in 1539.

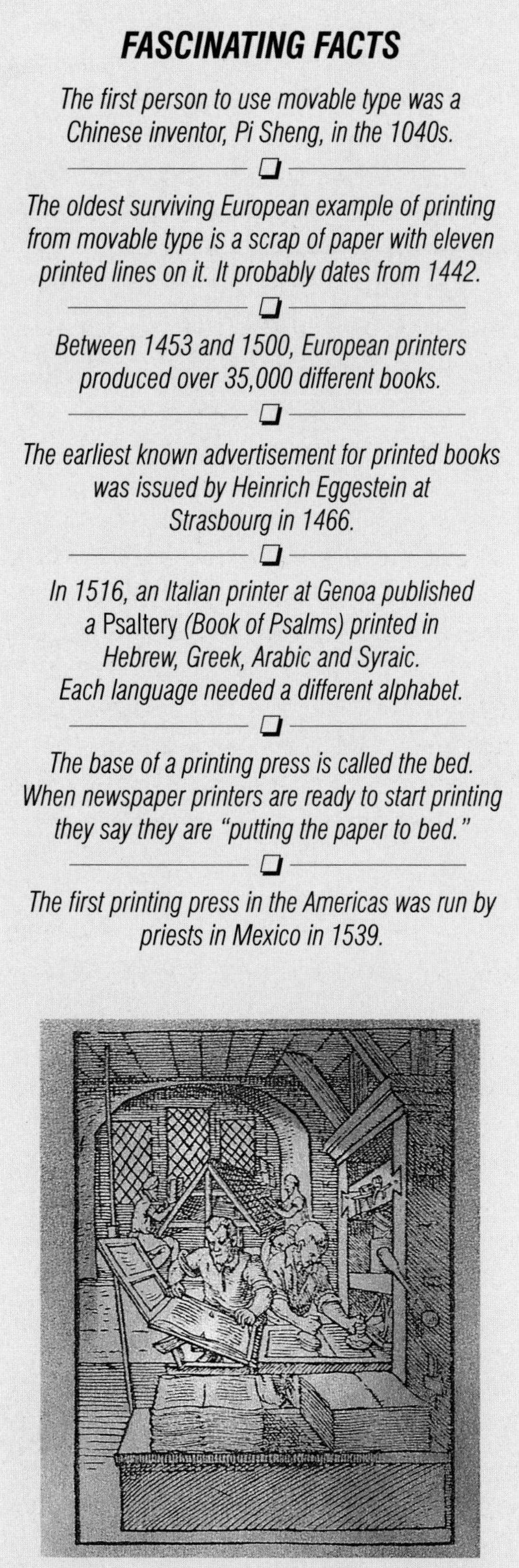

A printer's workshop from a sixteenth century woodcut.

A challenge to the church

The invention of printing from movable type happened at an important time in the history of European civilization. During the Middle Ages, the struggle to survive took up most people's attention, and they had little time for culture and the arts. The Christian church was the center of learning, but it had no interest in new ideas.

The fifteenth century brought a change, starting in Italy. There was a great upsurge of interest in painting, sculpture, architecture, music and literature. Historians call this change the Renaissance. In many ways, people were looking back to the days of Ancient Greece and Rome. But they also explored new ideas.

Another movement, the Reformation, affected the Christian church more directly. Until the fifteenth century, church services were sung in Latin, and priests read to the people from the Latin Bible. As only educated people understood Latin, what happened in church meant nothing to most of the people there.

Rushing into print

Once printing had been invented, there was no stopping it. Within a few years there were printing works at several cities along the Rhine, and then throughout Europe. The first press in Italy was set up in Rome in 1464. By the 1470s there were also presses in France, the Netherlands, Switzerland, Spain and England.

Among the first printed books were translations of the Bible into everyday languages. The first Bible in German was printed in 1466, and French, Spanish, Dutch and English

Casting letters by hand in molds. Some books and all newspapers were printed using individual metal letters until well into the 1970s.

translations soon followed. Suddenly many more people could read, and interpret, the Bible for themselves.

Soon, printed books on other subjects began to appear. The English printer William Caxton, who started his press in London in 1476, produced about 100 books before his death in 1491. They included favorite stories from literature such as *Reynard the Fox*, Chaucer's *The Canterbury Tales* and Sir Thomas Malory's *The Death of King Arthur* – and also a book about the rules and history of chess.

Telling the world

Printing a book was a major project, but printing one sheet of paper was fast and fairly cheap. Anyone with ideas could find an audience by having them printed and sending the sheets to friends. This alarmed the church, which was not used to being criticized, and also worried governments. The Roman Catholic Church, and even the Protestant churches of North America for a time, prohibited the printing of books about religion unless they had approved them first.

As for governments, they have always tried to control what is said in print, and continue to do so even today. But many printers defied churches and governments. Thousands of pamphlets (little booklets) on every topic of current interest were printed between 1520 and 1600. Wars as vicious as any experienced on a battlefield were fought by rival writers in their pages.

About 1600,the first real newspapers appeared. Often published weekly, they contained not only news, but also comments about public life.

Books for the people

For people who were not interested in politics, from about 1650 there were broadsides – single sheets with the words of popular songs. For family reading, there were chapbooks – little books containing one or two folktales, Bible stories or nursery rhymes. Then, in the 1800s, came cheaply printed editions of popular books – novels, biographies, science books and poetry.

Meanwhile, scholars and thinkers were able to publish their ideas in book form and could reach thousands of people instead of the few who could meet in a lecture hall. What is more, a book could be read anywhere, so ideas could take root thousands of miles away. In this way, printing made the world a smaller, as well as a more knowledgeable, place.

Galileo & His Telescope

Imagine seeing on the news this evening that scientists had proved that the world was flat! Everything we know about how the world works is suddenly revealed to be completely wrong.
That was the scale of the shock felt by the world when Galileo proved that the Earth went round the sun and was not the center of the universe.

1200 1300 1400 1500 1600 1700 1800 1900 2000

1610 Padua, Italy

Since the beginning of history, the sky has been the object of fascination and wonder. People from the earliest times had stared at the sun, moon, planets and stars and tried to make sense of their movements.

The Greeks get it wrong
The first scientist to come up with an answer was Hipparchus (Hip-ARK-kus), an astronomer (one who studies the stars) working on the island of Rhodes about 160–125 B.C.. His ideas were included in an encyclopedia of astronomy compiled by Claudius Ptolemy (A.D. 90–168) (TOE-lem-ee), who lived in Alexandria, Egypt.

Ptolemy said that the universe was like a hollow sphere, like a soccer ball, with the Earth fixed at its center. The sun, the moon and the planets all circled round the Earth. All the stars were fixed in place on the inside wall of the sphere; they never moved about.

Ptolemy's ideas about the universe were believed for 1,500 years. They were absolutely wrong.

Galileo Galilei (1564–1642) was born in Italy. He experimented with gravity by dropping balls from the Leaning Tower and noting how fast they fell.

Asking questions
About 1520, a Polish astronomer called Nicolaus Copernicus (1473–1543) began to question Ptolemy's ideas. He realized, from his observations of the sky, that the movements of the moon and planets did not fit in with Ptolemy's calculations.

As the telescope had not yet been invented, Copernicus could only make calculations based on his observations with the naked eye. These persuaded him that the Earth and the planets revolved round the sun, but he had no way of proving it.

Few other scientists were interested anyway. Ptolemy's explanation of how the universe worked seemed good enough to them. The Roman Catholic Church, which at that time controlled all the places of learning in Europe, wanted nothing to do with Copernicus's theory. In the Church's view, there was no doubt that the Earth was at the center of everything. This, the Church said, was made clear by a number of passages in the Bible.

Searching for proof

Anyone can have ideas. Proving them is more difficult. The scientist who proved that Copernicus was right was Galileo (GAH-lil-eh-oh), who was born in Pisa in northern Italy in 1564. He trained as a doctor, but became more interested in science and eventually became Professor of Mathematics at the University of Pisa.

Galileo investigated many scientific topics, including how fast objects fall and how a pendulum swings, but he became increasingly interested in astronomy. He had read Copernicus's book, *De Revolutionibus Orbium Coelestium* (On the Revolutions of Celestial Bodies), but was frustrated in his attempts to investigate further because he lacked a suitable instrument through which to observe the night sky. But then, in 1609, there was a breakthrough.

Improving the telescope

A traveller who had returned from the Netherlands brought with him a new Dutch invention, the telescope. It was a crude and not very effective instrument, but it inspired Galileo to try to improve it. Then, he thought, he

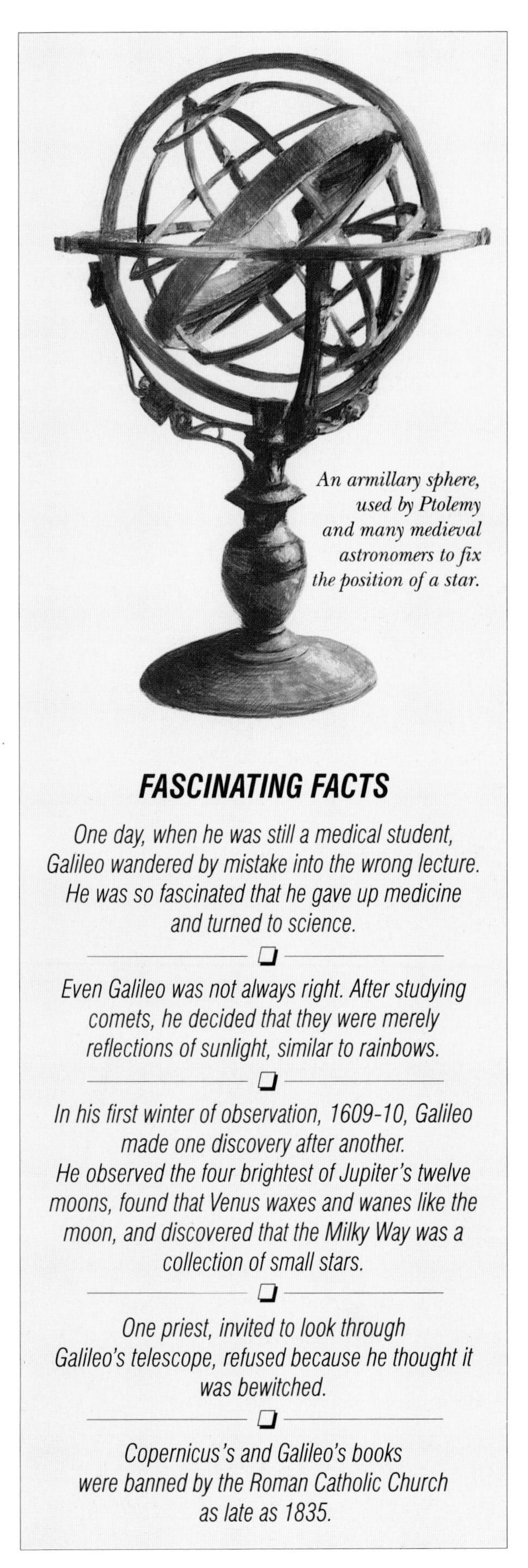

An armillary sphere, used by Ptolemy and many medieval astronomers to fix the position of a star.

FASCINATING FACTS

One day, when he was still a medical student, Galileo wandered by mistake into the wrong lecture. He was so fascinated that he gave up medicine and turned to science.

Even Galileo was not always right. After studying comets, he decided that they were merely reflections of sunlight, similar to rainbows.

In his first winter of observation, 1609-10, Galileo made one discovery after another. He observed the four brightest of Jupiter's twelve moons, found that Venus waxes and wanes like the moon, and discovered that the Milky Way was a collection of small stars.

One priest, invited to look through Galileo's telescope, refused because he thought it was bewitched.

Copernicus's and Galileo's books were banned by the Roman Catholic Church as late as 1835.

would be able to carry out proper observations and find out if Copernicus's theory was right.

Galileo worked out a way of grinding the glass lenses so that they gave a more accurate curve. He succeeded in making a telescope which could magnify objects thirty-two times. This was ten times better than earlier versions. Hundreds of copies of this telescope were sold all over Europe.

Galileo's original sketches of Saturn, Venus, Jupiter and Mars.

Dangerous ideas

Meanwhile, Galileo began his own observations. These soon confirmed the theory of Copernicus that the Earth travelled round the sun, as did the other planets. In 1613 Galileo published his findings in a series of newsletters known as *The Sunspot Letters.* At once, he was in trouble. According to the Church, the Bible agreed with Ptolemy's ideas, and Galileo's book was banned.

When he repeated his findings in a book written in Italian, published in 1633, he was tried by the Inquisition and condemned for the sin of heresy (contradicting the word of God). He escaped death only by denying his own theories and discoveries. It was reported that he muttered after he finished his denial. "Nevertheless, it does move!" – his final defiance of those who would deny scientific facts.

From then on, he lived quietly in retirement, but in 1638, four years before he died, he published a collection of all his scientific findings, which laid the foundations of the science of physics.

Pioneer in science

Galileo's work was a turning point in the history of science. His description of the universe was the starting point of modern astronomy. His discoveries in the area of physics called mechanics were the basis for the invention of the pendulum clock among other things.

Most important of all, his methods of working became a model for later scientists to follow. Galileo taught all scientists that the key to good work is tireless and repeated observation and experiment until a theory has been proved or disproved.

Alongside all this went another idea which was truly shocking to most people in Galileo's time. It was that what a religion teaches about the world is not always correct in reality. His astronomical work proved it. So he played a part, too, in the questioning of religious ideas which followed the split of the Christian Church into Roman Catholics and Protestants at the Reformation.

Galileo at work. On the desk is his leather-covered telescope. It is a tribute to his genius that he could make so many accurate observations with such a tiny instrument. The papers are proof pages of his first astronomical book, Siderius Nuncius *(Starry Messenger), published in 1610, which recorded his early observations of our solar system.*

SIDEREVS
GALILEO GALILEO

Newton & the Laws of Gravity

Why do apples fall to the ground? Why don't they fall upwards? What makes the tides go up and down? Why does a kicked soccer ball go faster than one rolling down a slope? Why do seat belts help in an accident? The answer to all these questions is gravity. Newton's pioneering work solved scientific mysteries and led to new inventions.

1200 1300 1400 1500 1600 1700 1800 1900 2000

1687 London, England

Isaac Newton was twenty-three and studying at Cambridge, when the Great Plague which had already caused panic in London reached the university city. As it happened in London, anyone who had somewhere to go headed for the country. Newton went back to his mother's farm in Lincolnshire where he had worked for four years before going up to Cambridge. He spent two years there until the fear of plague had died down.

An apple falls

Other students might have been frustrated by this interruption to their work, but Newton saw it as an opportunity to do some thinking. He became interested in the motion of objects. There is a story that his interest began when he watched an apple fall from a tree and wondered what force attracted it towards the ground. He went on to wonder what force it was that kept the moon close to the Earth instead of spinning off into space.

The foundation stone of his most famous work had been laid, but back in Cambridge he became interested in optics, the science of light. He made important discoveries in this field before returning to the force that he called by the Latin word *gravitas* (weight)and which we call gravity.

In 1687, Newton

Isaac Newton (1642–1727) read a paper on optics to the Royal Society in 1675. Called Discourse on Light and Color, *it included this diagram.*

One of Newton's earliest discoveries, when he was investigating optics, was that light shining through a prism splits up into the colors of the spectrum (the colors of a rainbow). The objects on his desk represent two other scientific interests in his life. The telescope is a reminder of his interest in astronomy, and the apple recalls his discovery of the force of gravity.

published his findings about gravity in a book called *Philosophiae Naturalis Principia Mathematica* (Mathematical Principles of Natural Philosophy). In it he suggested that there were three laws of motion which applied to objects on and around the Earth.

Laws of motion

His first law said that an object at rest will remain at rest unless an outside force acts on it. The second was that the movement of an object is always in proportion to the force applied to it. The third law was that for every action there is an equal and opposite action.

These three laws explained for the first time many common experiences. When a horse breaks into a gallop, the rider tends to be thrown backwards – this illustrates the first law. If you throw a ball hard, it travels faster than if it were tossed gently – an example of the second law. The third law is demonstrated when the force of a horse's hooves landing throws up a scatter of stones from a path.

Solving mysteries

Newton's laws solved many scientific mysteries, such as the motion of moons round the planets, the effect of the Earth's moon on the tides, and the behavior of comets. They also created an understanding of force, motion and acceleration which was to help the engineers who developed steam, and later internal combustion, engines.

In the twentieth century, some of Newton's ideas have been amended by the discoveries of Albert Einstein (see page 65–69) about the behavior of objects at very high speeds. But, in most circumstances, Newton's laws are still true.

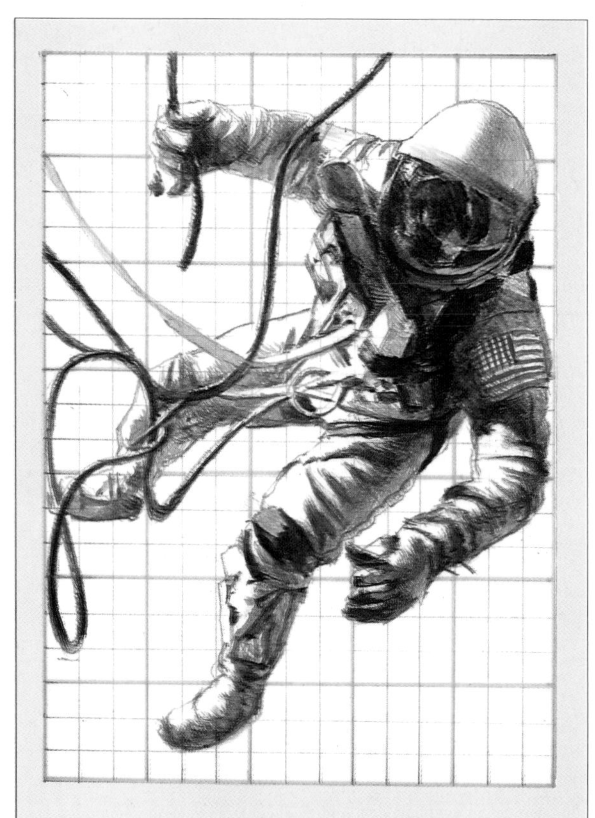

If there was no gravity, we would all float about like this space-walking astronaut.

FASCINATING FACTS

Newton made the first reflecting telescope in 1668, using a concave mirror to reflect the image. This idea is still used in today's large observatory telescopes.

❑

Newton taught himself mathematics. He invented calculus, which he called fluxions, about 1675. This is an essential mathematical tool for all scientists.

❑

Newton's Principia *almost failed to be published because of shortage of money. But his friend, the astronomer Edmund Halley, stepped in and paid for publication.*

❑

Newton was so respected by his fellow scientists that he was elected President of the Royal Society – England's leading scientific post – twenty-four years running, from 1703 until his death.

❑

The international unit of force, used by engineers, was named the newton in 1960 in honor of Sir Isaac Newton's achievements.

Watt & the Steam Engine

"I need a better power source," fumed Matthew Boulton. "One that doesn't get tired as horses do, one that works efficiently!" Then he met James Watt and the steam engine was born. Steam powered the machines that made the Industrial Revolution. Steam powered the railways and the ships – it totally changed how people lived, worked and travelled about.

1200 1300 1400 1500 1600 1700 1800 1900 2000

1769 Glasgow, Scotland

One day in 1764, a model engine was brought in to the workshop of a Glasgow instrument-maker for repair. It was a miniature Newcomen engine, part of Glasgow University's collection of scientific exhibits. The workshop belonged to James Watt, then 28 and the scientific adviser to the University.

One-stroke wonder

The Newcomen engine had been invented about fifty years earlier to pump water out of mine-shafts. It was a great improvement on an earlier steam-operated pump, and was used in hundreds of coal and other mines. But it was inefficient. It used huge amounts of fuel, and it worked jerkily.

These were not serious problems when it was used for pumping water from coal-mines – there was plenty of cheap coal to be had, and no one worried if a pump was jerky. But other possible users, who did not have a supply of cheap coal, or who needed an engine that would run evenly, were not interested in the design.

James Watt (1736–1819) made a breakthrough with his first model of a steam engine with a separate condenser.

Asking questions

James Watt got the University's model engine running again. It was the first Newcomen engine he had seen and he studied it with interest. Why was it so inefficient? Why was the motion so jerky? Why did it use so much coal?

The Newcomen was a one-cylinder engine. The cylinder had a piston inside it, connected to a beam that operated the pump. Steam from a boiler entered the cylinder from below and forced the piston up, operating the beam. Then cold water was forced into the cylinder to condense the steam so that the piston dropped down again.

Each time the cold water was forced in, the steam already in the piston and the fuel which had produced that steam were immediately wasted. New steam had to be generated in the boiler, using more fuel, for the next stroke of the piston.

Cutting fuel costs

Watt's solution was to add a second cylinder, called a condenser, with a connection between the two. The waste steam was turned back into water in the second cylinder, so the first cylinder could be kept hot all the time, with a great saving of fuel. In fact, Watt's engine used only a quarter of the fuel of the older one.

Watt's idea did not bring him instant fame and fortune. Like many other inventors, Watt could not find money to back him. It was another twelve years before the first of his engines went on sale. In 1776 that he formed a partnership with a Birmingham manufacturer, Matthew Boulton (1728–1809), to make his engine.

This was only the beginning. Watt continued to work on more improvements and eventually came up with an engine that, instead of producing an "up and down" or "reciprocating" movement, made a "round and round" or "rotatory" one.

James Watt was a scientist and inventor, not a businessman. Without the interest of Matthew Boulton, Watt's ideas would probably still be locked up in his notebooks. Helped by Boulton, the condensor was only the first of Watt's many contributions to the technology of steam.
The Industrial Revolution brought together inventors with ideas and business people who had the ability to translate those ideas into money-making ventures.

Steam on the move

This was the real breakthrough. A "rotatory" – or, as we would say today, rotary – engine could be used for a far wider range of purposes. It could drive rotating shafts which, connected with canvas belts, could operate machinery in cotton mills and other factories. It could also drive the wheels of vehicles.

The first experiments with steam-driven vehicles took place in the 1770s. These early vehicles ran on the roads, but it was not long before engineers were thinking of another application of the steam engine, as the power source for the first railways. Watt, who died in 1819, lived just long enough to see the start of the Industrial Revolution which his work had created, but missed the beginnings of the Railway Age by about six years.

The Steam Age

By the middle of the nineteenth century the industrialized world was "run by steam." Steam powered the machines in factories making anything from cloth to cannons. Steam hauled the trains and even powered the equipment that built the railways. At sea, steam was taking over from sail.

But the Industrial Revolution was not only a matter of new technology, important though this was. New machines and new factories to house them cost money, most of which had to be paid long before the factories were producing goods which could be sold at a profit. Just as Watt needed Boulton to back his ideas with investment, so as the Industrial Revolution got under way, investors' money was needed to help it succeed.

The Industrial Revolution was also a Business Revolution, with investors

FASCINATING FACTS

The first practical machine powered by steam, was the Miner's Friend. It was patented in 1698 by Thomas Savery, a Cornish mining engineer.

❑

Hurrying to make a model of his first engine, Watt stole a thimble from his wife's sewing-basket to make a vital stop for the end of a pipe.

❑

Before he met and went into partnership with Matthew Boulton, Watt became so disheartened that he gave up steam and got involved in building canals in Scotland for four years.

❑

Watt was not only an improver of steam engines. He also invented a way of copying documents by a chemical process, which became part of every office's equipment for 100 years.

❑

In 1784, Watt heated his offices with steam pipes, the first time steam was used for central heating.

❑

In 1789, Watt invented the governor, a device that controls the speed of a steam engine. This allowed the engine to run at a constant speed, regardless of the heat of the boiler, and so drive other machines reliably.

❑

The unit of power, the watt, was named for James Watt during the nineteenth century.

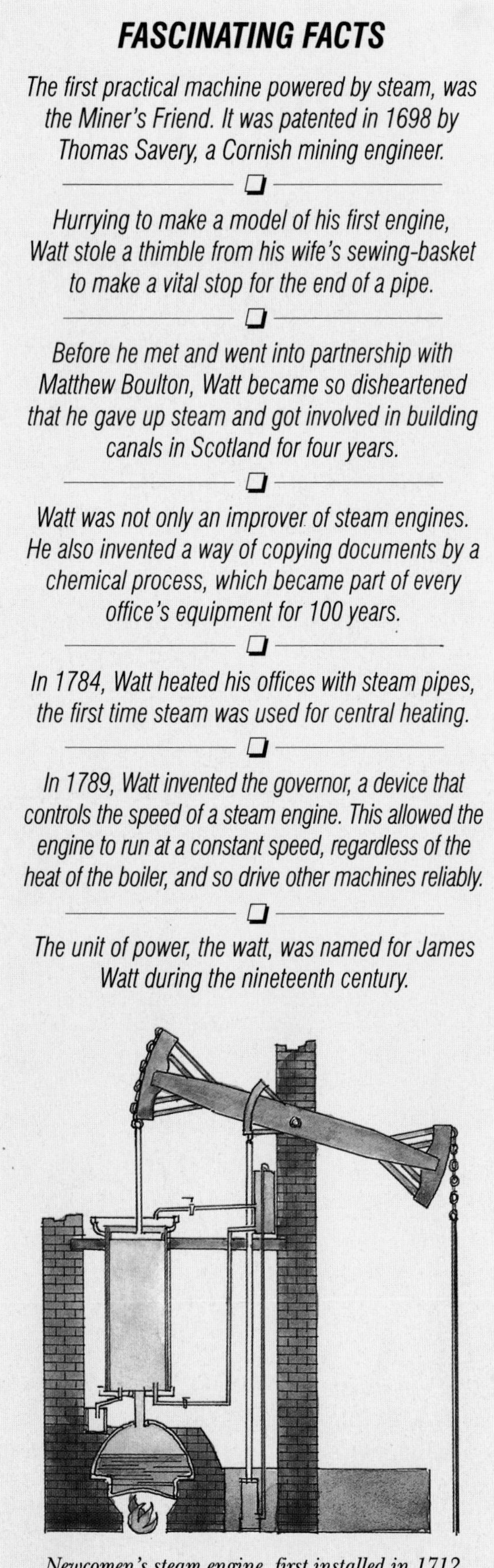

Newcomen's steam engine, first installed in 1712.

Joseph Cugnot, a French military engineer, built this steam carriage around 1771. It could carry four people at speeds of up to 2.5 mph.

meeting to form companies, banks lending money to industry, and business people making a living out of bringing ideas and money together.

Winners and losers

What about the people who were not inventors or investors or bankers or manufacturers? The steam engine changed many of their lives too. In the eighteenth century, weaving, for example, was a hand- or foot-operated craft carried on in small workshops, often as a family business. The new mills, where cloth was woven by steam-powered looms, employed hundreds of people – including very young children – performing simple and repetitive tasks. Craftsmen had been replaced by machines.

Factories created millions of new jobs – but they were not as pleasant or as satisfying as craft work. They also meant moving to industrial towns where housing was cramped and the air – with factory chimneys belching out pollution from their steam engines – was foul. Meanwhile, the demand for coal to fuel the Industrial Revolution created thousands of jobs in deeper and more dangerous mines.

After steam

The Steam Age ended in the 1950s – it had lasted for just over 150 years. Industry today is powered by electricity. For transport we rely on both electricity and oil and there are no more steamships. But the huge changes brought about over the past two centuries were begun by steam, and by James Watt.

Not all of those changes have been for the better. Rivalry between the world's industrial powers have led in this century to two World Wars. There is still a great gap between the wealth of industrialized countries and the rest of the world – millions of people who work in factories, wish there was some other way of earning a living. When James Watt looked at Newcomen's engine and thought about his improvement, he started something that changed the world.

Preserving Food

"Thirsty? Have a can of Coke...." "Hungry? How about some soup? Let's open a can...." Before the invention of canning, the only way to preserve food was to dry it or salt it. Both processes took time and the end product needed a lot done to it before it could be eaten. With cans, suddenly people could eat good food fast in the Arctic, jungle or desert.

1200 1300 1400 1500 1600 1700 1800 1900 2000

1813 London, England

You have come home late after a long, tiring day. You are too tired to do much cooking. What you need, before you go to bed, is something tasty and quick to prepare. You look at the cans in the kitchen cupboard – there are baked beans, tomato soup, tuna, peaches....

Food for battle

Almost every kitchen today will contain a stock of canned foods. They provide quick snacks, meals for unexpected visitors, and fruit and vegetables out of season. But the food canning industry did not begin as a convenience for people at home. It had a more serious aim – to provide wholesome food for soldiers in battle.

The story began in France in 1795, when the Emperor Napoleon Bonaparte offered a prize for a practical method of preserving food.

In 1804, a Paris chef, Nicolas Appert (1752–1841), discovered that liquids such as soups and stews, and small fruits like raspberries and cherries, could be preserved by packing them into champagne bottles and then plunging the sealed bottles in a bath of boiling water. He did not know why this kept the food fresh. In fact, he was sterilizing it, killing the bacteria which make food go bad. It would be over fifty years before the French scientist Louis Pasteur (1822–95) made the discoveries about bacteria which gave the explanation.

Hall and Donkin presented cans of beef to George III in 1812. Word came back that the King had found the contents "most palatable."

British enterprise

By 1807, Appert was supplying the French

A scene in an early canning factory: the man in the foreground is soldering lids on the cans, while a boy heats up a soldering iron at the stove.

army and navy with soup, peas and beans in bottles. He published a book about his work in 1810, which was seen by two British industrialists, John Hall and Bryan Donkin. They bought the English rights to Appert's patent.

The neck of a bottle limited the kinds of foods which could be preserved in this way, so Hall and Donkin experimented with cans made of tin-plate – sheet iron covered with a thin film of noncorrosive tin. Their first product was canned beef, and by 1818 they were supplying tens of thousands of cans to the British navy.

Canning by hand

At first, the cans were made and filled by hand and sealed by soldering the lids in place. This made them too expensive for ordinary people to buy, but canned food became very popular with soldiers, sailors and explorers.

It was in the U.S. that food canning first developed into a mass production operation with goods that most people could afford. About 1890, the U.S. became the world's leading producer and consumer of canned food.

Ready for use

Canned food was the first type of "convenience food" which could be stored for long periods in the kitchen ready for when it was needed. It released cooks from many hours of work and also enabled them to make more interesting dishes using canned ingredients. It gave people a more varied diet, and gave a huge boost to the fishing industry and to farmers producing fruit, vegetables and meat for canning. More recently, aluminum cans have provided a convenient and lightweight way to buy drinks.

FASCINATING FACTS

The first food cans had to be opened with a hammer and chisel. The key-opening can was invented by an American, J. Ousterhoudt in 1866. The ring-pull drinks can was invented in 1978.

In 1830, a can of salmon cost as much as most people could earn in a day, so it was a luxury that only the rich could afford.

Canned food spelled disaster for explorer Sir John Franklin's expedition to the Arctic in 1845. Not one of the 129 officers and men survived, partly because the canned food they had taken with them had gone bad.

The British navy took canned food on expeditions to the Arctic in 1814

The first sweetened, condensed milk was canned at Torrington, Connecticut, in 1856 by Gail Borden. He also invented powdered milk.

Cans taken to the Antarctic by Captain Scott's expedition in 1912 were still intact when one of his food dumps was found in the 1960s.

Heinz has been selling baked beans in Great Britain since 1895; they estimate that over 547,500,000 cans of baked beans were sold there in 1992.

The Invention of Photography

Imagine only knowing what something or someone looked like if you had actually met them or seen a painting. Nowadays we all know what the Eiffel Tower looks like, and we would recognize Madonna though we've never met her. Photos have done this; and they also provide a record of events that happen in seconds, like a disputed goal or a bank robbery.

1200 1300 1400 1500 1600 1700 1800 1900 2000

1839 Paris, France

When you take a photo, the light rays from the subject of the picture pass through the lens of the camera and focus on the film. The chemicals on the film store the image, which can then be developed and printed.

The behavior of light when it passes through a small hole was known by Arab astronomers 1,000 years ago. They used a camera obscura to observe eclipses of the sun. This was a lightproof tent, or room, with a small hole in the roof through which light could pass. This light produced an upside-down image that was displayed on the wall. Later, cameras obscura were used by artists, who copied the images, and as tourist attractions. But no one had found a way of fixing the image permanently.

Chemical magic

It was clear that if the image could be fixed permanently, it would be done by treating the surface on which the image fell with chemicals. Many inventors succeeded in doing this, using glass, metal, or paper coated with various chemicals. The first was Louis Daguerre (1789–1851), who succeeded in fixing a permanent image on a metal plate in 1839.

By the 1860s photographers were taking high-quality pictures on chemically coated glass plates. Their subjects included portraits, scenery, and news events. In 1875 the halftone process was invented by Karl Kleitsch. This turned photos into dots so that they could be reproduced in print.

Joseph-Nicéphore Niepce (1765–1833) produced the world's first real photograph on a metal plate in 1826.

Pictures for the people

The next important step was the development of roll film by George Eastman (1854–1932), replacing glass plates, and this was followed in 1888 by the first simple box camera designed for amateurs to use. The first roll films were made of paper, but

FASCINATING FACTS

The Englishman Thomas Wedgwood was probably the first person to find out how to take photographs – but they faded before his eyes.

❑

The first book illustrated with photographs was The Pencil of Nature, *published by William Henry Fox Talbot in 1844. It included pictures of his family and servants.*

❑

Metal-plate photographs, called daguerreotypes after their inventor, the French scientist Louis-Jacques-Mandé Daguerre, were unique. You could not make a copy of them.

❑

The first leading woman photographer was English; her name was Julia Margaret Cameron (1815–79). In 1864 she started taking portraits of literary and theatrical personalities.

The American Civil War was the first war to be properly recorded in photographs.

in 1889 Eastman introduced flexible plastic film. This opened the way for the invention of cinematography and the development of the motion picture industry.

The first full-color photography process was invented in France in 1903, although it was not until the 1960s that color film became widely available for amateur photographers.

Finally, in 1947 the American scientist Edwin Land invented the Polaroid instant camera, which develops and prints the image within a few seconds of taking the picture.

Family records

Photography is one of those inventions whose effects have spread into almost every area of life. It is difficult to believe that until about 150 years ago, there was no way of knowing what someone looked like unless you had met them. The best you could do was to have an artist paint a portrait.

Unless you took an artist with you, or made sketches of your own, you could not record your holiday memories. Parents could not record pictures of their children as they grew up. But the enrichment of our personal lives is only one benefit of photography.

Serving science and history

Thanks to the camera, we can now study in detail the events and people of the past 150 years. The events of the American Civil War (1861–65), for

William Henry Fox Talbot (1800–77) was an English squire whose hobby was photography. His first real photograph on chemically treated paper in 1839 showed a diamond-paned window. Later, he discovered a way of making contact prints, or copies, of his pictures.

example, were recorded in hundreds of photographs. We know exactly what the heroes and villains of the recent past looked like, from Martin Luther King, Jr., to Adolf Hitler. In tomorrow's newspapers, we will see photographs of events that took place today on the other side of the world.

Photography has also revolutionized science. Scientists can record and check their scientific processes and can photograph tiny objects under the microscope. Astronomers record their observations on film so that they can study them in detail later. Pictures taken from space have increased our knowledge of the solar system.

Saving lives

One of the most important and life-saving uses of photography is called radiography – the photography of organs inside the body using X rays. X rays were discovered in 1896 by the German scientist Wilhelm Konrad von Roentgen (1845–1923) (RONT-gen). X ray photographs show tissues deep inside the body, and by studying the pictures doctors can tell whether our body organs are diseased or healthy. Dentists also use X rays to study the condition of our teeth and decide what treatment is needed.

Perhaps the most important result of the invention of photography is the effect it has had on the way we look at the world. We are no longer happy just to read or be told about something. We want to see a picture of it. In the late twentieth century we have gotten used to looking at photographs for information, education, and pleasure.

The first daguerreotype photographs required that subjects sit without moving for up to five minutes. Professional studios used neck braces and other props to help clients do this.

Darwin's Theory of Evolution

"Men descended from monkeys!" – "Bible account a lie!" These were the type of outraged headlines that greeted Darwin's theory of evolution in 1859. Good Christians were horrified at the unstated attack on the Bible. Others felt threatened by the idea that humans and animals had the same ancestors – their feeling of superiority over animals was in question.

1200 1300 1400 1500 1600 1700 1800 1900 2000

1859 Galapagos Islands

The year was 1831. Charles Darwin, a student at Cambridge University, had been offered a place as naturalist on a five-year, around-the-world, scientific voyage aboard a ship of the British navy, the HMS *Beagle*.

New horizons

It was the chance of a lifetime. The HMS *Beagle* planned to make a survey of the South American coast and the islands of the eastern Pacific. Darwin would have plenty of opportunities to spend time ashore while the survey work continued.

For those five years, Darwin worked ceaselessly. He filled his notebooks with descriptions and sketches of plants, fossils, amphibians, mammals, and birds. He collected hundreds of specimens, which were carefully packed into boxes for the voyage home.

Charles Darwin (1809–82) was twenty-two when he began the voyage on the Beagle. *He had already studied medicine and theology before turning to biology.*

13 different finches?

It was in the Galápagos Islands, off the coast of Equador, that Darwin first noticed something strange. Each island seemed to have a different species of finch, although some of the islands were so close that you could see the one next door.

When Darwin returned to England in 1836, he began the long task of writing up his specimens and making sense of what he had seen. One of the specialists who helped him, John Gould, confirmed that the Galápagos finches were all different species. Each species had a different-shaped beak. This meant that they could not breed with each other and were unique to one particular island.

The great idea

Darwin ended his five-year expedition with one overriding impression from the many experiences it had provided. It was that the differences he had observed between living things of the same or similar species did not fit in with the current scientific theory (based on the stories in the Book of Genesis in the Bible) that species died out and were replaced by new ones. He was sure that things were not as simple as that.

He was not the first to question whether the Bible was right. His grandfather, another famous naturalist named Erasmus Darwin (1731–1802), had also written of his doubts. So had an eminent French naturalist, the Chevalier de Lamarck (1744–1829).

Can a species change?

By 1842, Darwin had worked out the basis of an explanation. He called it natural selection. All the individual members of a species at any one time are not identical. For example, some of the birds of a particular species might have longer beaks than others. These longer beaks might give them an advantage in obtaining food, and so they would have a better chance of survival than shorter-beaked birds.

In time, as the long-beaks bred, they would become a new species in which long beaks were normal, while the short-beaks would shrink in numbers and possibly die out altogether. The survivors would be the birds best fitted to their environment.

Changes like this, Darwin said, had

In the Galápagos Islands in the Pacific, Darwin studied the different species of finch of the group on the thirteen islands. They seemed to have evolved differently according to the particular conditions on their island. They were to become an important example in his discussion of evolution.

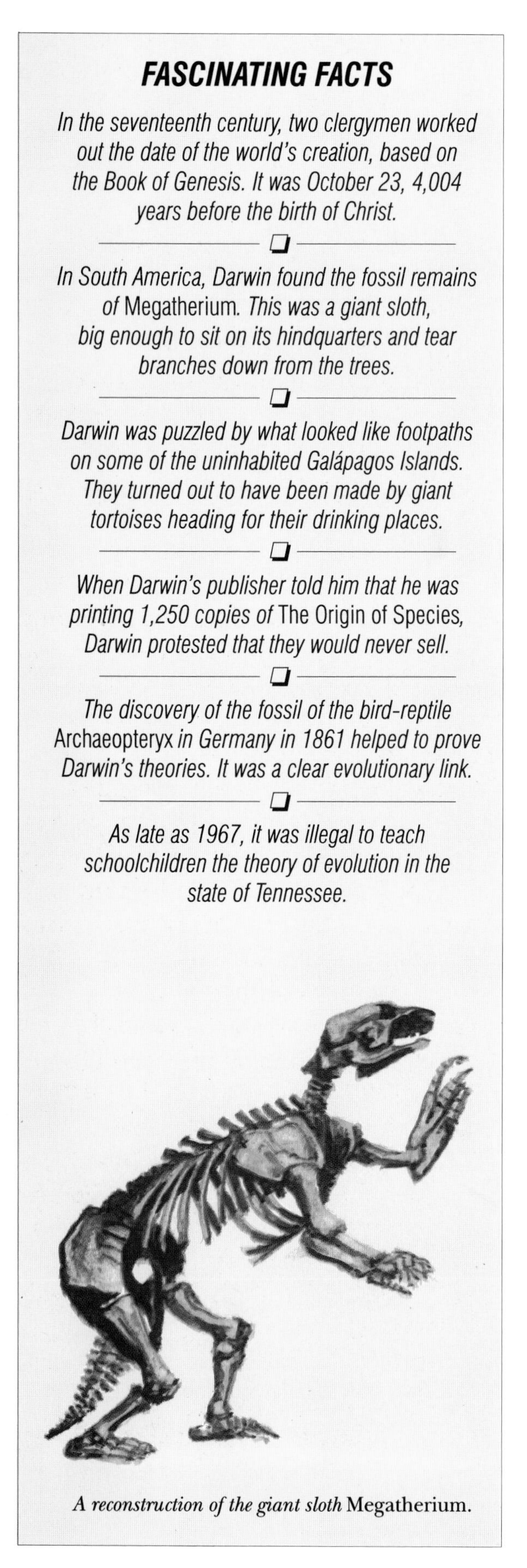

FASCINATING FACTS

In the seventeenth century, two clergymen worked out the date of the world's creation, based on the Book of Genesis. It was October 23, 4,004 years before the birth of Christ.

In South America, Darwin found the fossil remains of Megatherium. *This was a giant sloth, big enough to sit on its hindquarters and tear branches down from the trees.*

Darwin was puzzled by what looked like footpaths on some of the uninhabited Galápagos Islands. They turned out to have been made by giant tortoises heading for their drinking places.

When Darwin's publisher told him that he was printing 1,250 copies of The Origin of Species, *Darwin protested that they would never sell.*

The discovery of the fossil of the bird-reptile Archaeopteryx *in Germany in 1861 helped to prove Darwin's theories. It was a clear evolutionary link.*

As late as 1967, it was illegal to teach schoolchildren the theory of evolution in the state of Tennessee.

A reconstruction of the giant sloth Megatherium.

affected all the plant and animal species now on earth – including humans – and had taken place over millions of years.

Trouble ahead

The observations that Darwin had made around the world suggested that his theory was right, but he was not content with that. He spent many years discussing these observations with leading scientists, animal breeders, and gardeners, and set up a pigeon-breeding program of his own.

This was not merely a search for scientific accuracy. He knew that once his ideas were in print he would be in trouble. For behind his theory lay a challenge to what the Bible said about how the earth was created in six days.

When he eventually published his great work, *The Origin of Species by Means of Natural Selection*, in 1859, he was careful to stick to scientific observation and theory about animals and not to add comments about humans that would upset the Church.

A bishop on the attack

It was not just the suggested time scale of evolution that shocked Christians. Darwin also theorized that all animals, and this included humans, were descended from four or five original creatures, and all plants from a similar number. The Bible taught that human beings were made "in the image of God." One bishop wrote that Darwin's theory was a "degrading notion."

Darwin had not said that humans were the descendants of apes – but that was what many people pretended to believe, in order to attack him. But that ducked the real issue. There were now two views of how the world was made and how human beings fit into it – the Bible's and Darwin's. If Darwin was right, then the Church had been misleading people for over 1,800 years.

There were also objections to *The Origin of Species* from scientists who realized that it conflicted with their own work. But many other scientists backed Darwin and found that his theories agreed with what they had discovered themselves.

Darwin's microscope. He used this to study insects and plants during his voyage on the Beagle.

Another storm

Encouraged by this, Darwin went on to publish *The Descent of Man and Selection in Relation to Sex* in 1871. This suggested, more directly than his earlier book, that humans and apes shared a common ancestor millions of years ago. There was another enormous fuss, but by now Darwin's reputation was secure and an increasing number of scientists all over the world had accepted his ideas.

Darwin's theory was not the last word on evolution. We shall probably never know exactly how the world's plant and animal species evolved. But more recent studies by biologists – in heredity and genetics (see pages 79–80) – have fit in with Darwin's theory, sometimes extending or modifying it.

Darwin altered the way that people look at the natural world. Part of our concern about the damage to the environment caused by human activities, for example, is due to our fear that it may lead to undesirable forms of evolution. Doctors' understanding of some diseases and disabilities that are passed on from one generation to the next has been helped by the work Darwin started.

Peace with the Church

Meanwhile, what happened to the Church's version? There are still some Christians who believe that the story of the creation in the Bible is literally true. But most accept that the creation story in the Book of Genesis is a mythical version of the truth.

In a strange way, Darwin and the Church made their peace when he died in 1882. He was buried in the place dedicated to Britain's great scientific, artistic, and military heroes – in Westminster Abbey, London.

The coelacanth was thought to be extinct until 1938, when one was caught off East Africa.

Lister the Germ-killer

Is it better to be in pain or dead? That was the choice posed by surgeons during the early nineteenth century, when every third patient who underwent an operation died from an infection gained during that operation. Without Lister's pioneering work on antiseptics, there would be no organ transplants, no open-heart surgery, no hip replacements...

1200 1300 1400 1500 1600 1700 1800 1900 2000

1867 Edinburgh, Scotland

Our skin is a protective case which stops germs entering our blood stream. That is why, if we cut or scratch ourselves, we cover up the tear in our skin with antiseptic cream and a bandaid. When the skin has healed, we can take the bandaid off.

When surgeons carry out an operation, they have to make cuts in our skin to remove or repair damaged or diseased parts of our bodies. Unless they take care, there is a danger that wounds will become infected during the operation or afterwards, before the cut in your skin has healed up.

Life or death

Up to less than 150 years ago, there was a real risk that even if an operation were successful, the patient might die from infection. The result was that operations were carried out only if the patient would otherwise die anyway. Operations merely to relieve pain were thought much too risky. It was far better, surgeons said, to be in pain than to run the risk of being dead.

Surgeons 150 years ago worried about the number of patients who died of infection, but they did not understand why their patients were dying. Then, in 1860, Joseph Lister, who was the Professor of Surgery at Glasgow University in Scotland, began to think about the problem seriously.

Joseph Lister (1827–1912) pioneered using antiseptics in surgery, making it a safe and routine form of hospital treatment.

Danger in the air?

Lister had been reading the reports of Louis Pasteur (1822–95), the French chemist, who had written a book about the effects of micro-organisms – germs – in the air. Pasteur had discovered that germs existed while investigating what was causing beer to go bad. Lister wondered if they were the problem? If germs could be kept out of the operating room, and away from open wounds, could infection be avoided?

Carbolic killer

Lister decided that he would try the experiment. He talked with Crace Calvert (1819–73), the Professor of Chemistry at Manchester University, who had been trying out acids that would prevent corpses from rotting. Calvert recommended an antiseptic liquid called phenol – or carbolic acid – to keep wounds clean.

In 1865, Lister began using carbolic acid in his operations. At first, he was unsuccessful, but soon he was able to show that dressing wounds with carbolic acid prevented infection and aided healing. Using this method, he was able to set and save the badly broken arms or legs of accident victims which would otherwise have had to be amputated (cut off).

He also discovered that bed sores and other already infected wounds could be cured by using antiseptic dressings on them. He published his findings in a British medical journal called *The Lancet* in 1867.

Saving life worldwide

Some surgeons were critical of Lister. How dare he say that their hospitals were full of germs? But surgeons using Lister's methods soon reported success from the USA, Germany, Russia, Austria and Denmark. Dr. George Derby of Boston, Massachusetts, was using Lister's techniques in his operations in late 1867.

From dressings, Lister moved on to the question of making the air in

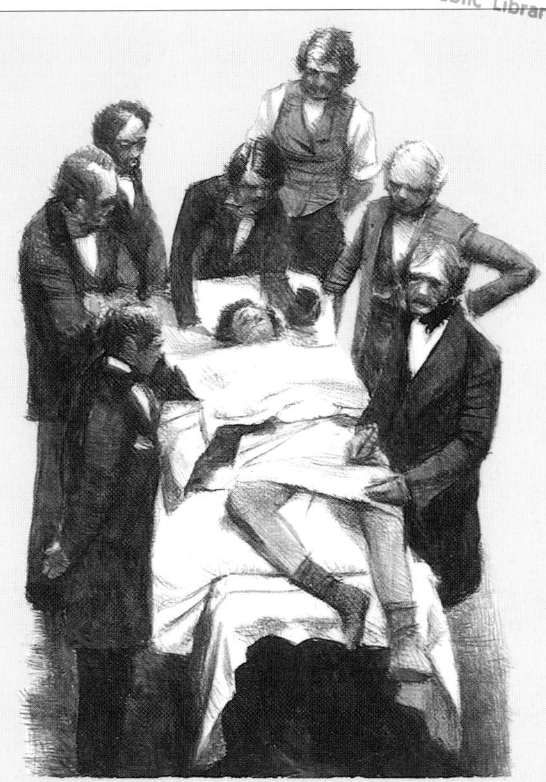

Lister had an early success with carbolic acid while setting the badly broken leg of a foundryman called John Hainy.

FASCINATING FACTS

An opinion poll of leading London surgeons in 1869 found that most of them thought Lister's work was "quite useless" or "meddlesome."

Surgeons first used rubber gloves in 1890. Face-masks first appeared in 1899.

Appendicitis used to be a fatal illness because the operation to remove the appendix – part of the digestive system – was too risky.

The first open-heart operation was carried out by an American surgeon, Daniel Williams, in 1893.

The first total artificial hip replacement unit, made out of stainless steel, was developed in England in 1938 by Philip Wiles.

The first human heart transplant was carried out in South Africa in 1967 by Professor Christiaan Barnard. Louis Washkansky, who received the heart, lived for 18 days after the operation.

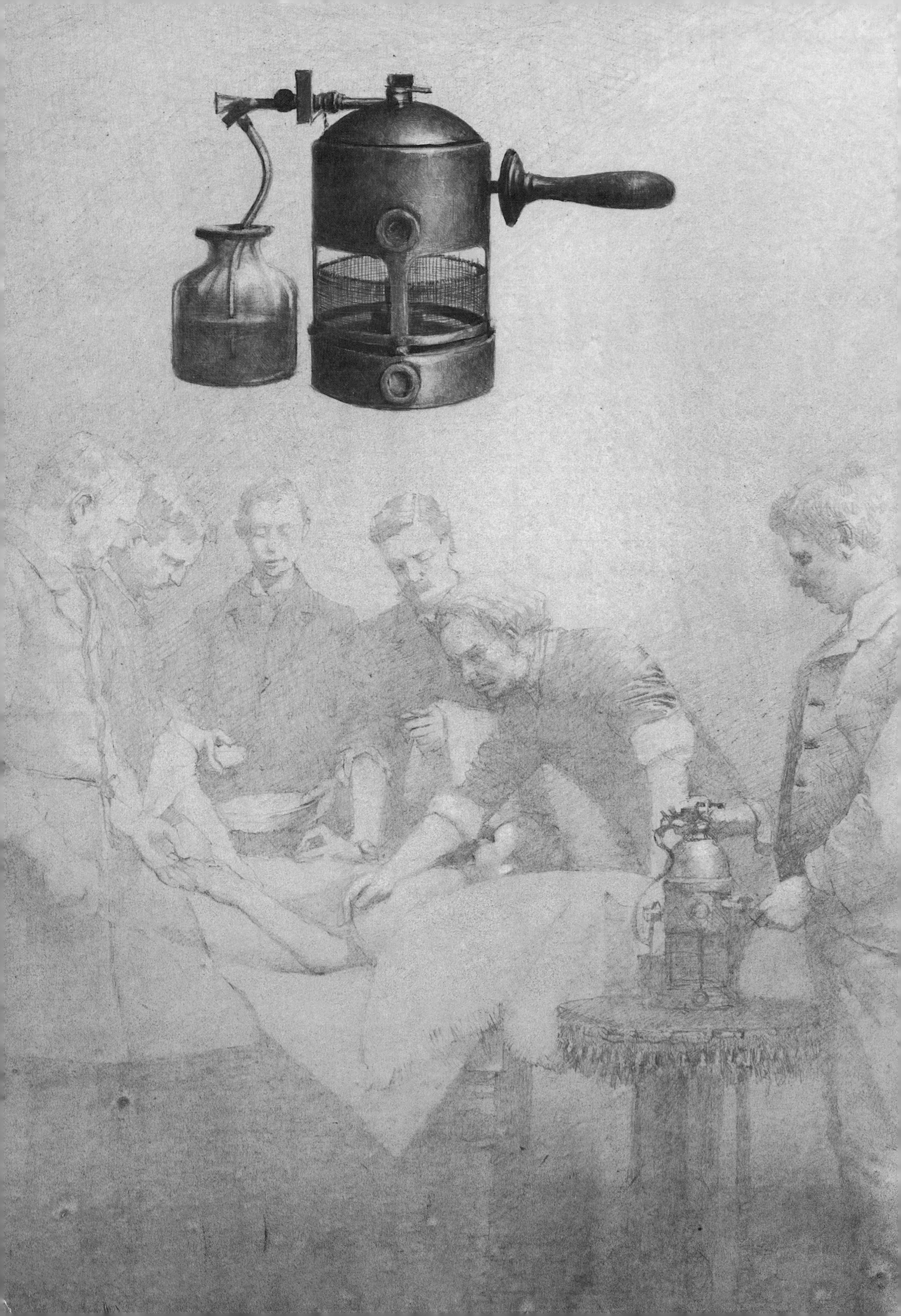

operating rooms germ-free. He developed a hand spray that could pump carbolic acid diluted with water into the air during operations.

Lister went on to work on asepsis – avoiding infection by making the operating room, the surgeon's hands, clothes and instruments all completely germ-free.

Louis Pasteur (1822–95) discovered how germs work. He developed vaccines against cholera, the plague, rabies and other dangerous diseases.

Meanwhile, the death rate as a result of surgery fell from one in three to about one in twenty. As operations became less risky, surgeons were more willing to perform them, and patients were more willing to undergo them. The number of illnesses that could be cured by an operation increased. The success rate increased the confidence of both surgeons and patients.

Improved techniques

There have been many improvements in antiseptic techniques since the 1860s. Lister himself abandoned the antiseptic spray in 1887 when he came to realize that the air in the operating room was less dangerous and germ-laden than he had thought.

Nowadays, air enters an operating room through filters which remove any harmful organisms. Instruments used in operations are sterilized – made germ-free – by placing them in boiling steam or under radiation. Surgeons wear sterilized masks, gowns and gloves, and materials for an operation are stored in sealed, sterilized packs until they are used.

The use of antiseptics has spread far beyond hospitals. Every pharmacist stocks antiseptic creams and lotions for home use. Training in their use is given to doctors, dentists, nurses, ambulance staff and first-aid workers. Food poisoning is less common because cooks and restaurant staff are aware of how germs can infect what they cook.

This scene in an operating room of Lister's time would horrify modern surgeons. The onlookers are wearing their everyday clothes which are full of germs. The surgeon and his assistants have no masks or gloves. But the picture also shows the beginnings of modern surgery. The man on the right is using an antiseptic spray, while an assistant is ready with an antiseptic swab.

Modern medicine

Surgeons today have developed their skills in ways that Lister could never have imagined. They transplant hearts and kidneys from one human to another and they replace the corneas (lenses) of eyes so that people with cataracts can see again. Surgeons can operate on babies before they are born or cut cancers out of the human brain. Blood lost in accidents or operations can be replaced by transfusions.

None of this would be possible without the use of antiseptics. Surgery has come a long way since Lister first started dressing wounds with carbolic acid. But because of his determination to overcome the problem of patients dying from their infected wounds, thousands of operations are now carried out daily in complete safety.

Bell & the Telephone

There is no communication as instant as the telephone. On radio and TV, events are described as they happen over a phone line. Push a few buttons and you can be talking to someone in Germany, or Australia, or Hawaii, or your grandmother in Miami. But back in 1876, when Bell made his invention, you could only send them a letter or a telegram....

1200 1300 1400 1500 1600 1700 1800 1900 2000

1877 Boston, Massachusetts, USA

"What did I say?"

"You said, 'Mr Watson, come here, I want to see you.'"

The two men, Tom Watson and Alexander Graham Bell, stared at each other. They had done it! Their telephone worked!

End of a struggle

The date was March 10, 1876. The place was Boston, Massachusetts. Watson had just rushed in from the next room to tell Bell that he had heard his voice down the wire. Their long struggle was over. They had just succeeded in sending an understandable message along a wire to another receiver.

It had been a worrying time. Bell had been working on the idea in his spare time for three years. He knew that a rival, Elisha Gray (1835–1901), an expert on telegraphic machines, was working along the same lines and had the backing of the Western Union Telegraph Company. Bell had only two scruffy rooms over a restaurant and one assistant.

Then, a few days earlier, he had heard that the great inventor, Thomas Alva Edison (1847–1931), was also in the race with a whole research laboratory of his own.

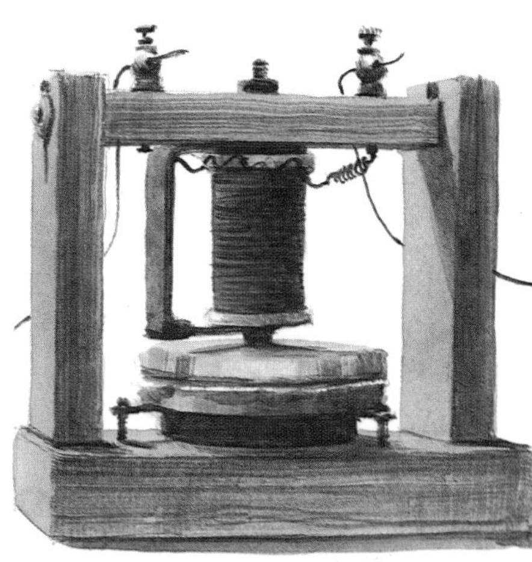

Alexander Graham Bell (1847–1922) built his first telephone using odds and ends from an electrical shop and the skill of his assistant, Tom Watson.

Stumbling into fame

The interest of Bell's rivals in the telephone had sprung from their knowledge of electricity. His came from his work as a teacher of the deaf. This had given him an understanding of sound vibrations, and he had stumbled on a way of changing these vibrations into electrical signals. In turn, this had led to his telephone experiments.

Bell's family had emigrated from

Scotland to Canada when he was twenty-three, and soon afterwards he had moved to Boston. That was six years before. Now, still under thirty, he was at the gates of fame and fortune.

Start-up problems

By July 1876, the Bell Telephone Company was in business. But there was trouble ahead. There were legal disputes with Bell's rival inventors, and although he won these battles, they were a worry. Bell also had to struggle to get his telephone accepted. Then there were technical problems, such as the organization of exchanges so that telephones could be connected.

But by 1887 there were over 150,000 telephones in use in the United States, and about half that number in Europe. From then on, the spread of telephone networks across the world never let up.

Buzzes and squeaks

Although the lines often crackled and spluttered, and speech was distorted, the first telephones seemed to most people like a miracle. It was almost unbelievable that you could have a conversation with someone in another city, or even in another country. We are so used to this idea today that it is difficult to imagine how amazing it was in 1876.

Because Bell was so young when he made his breakthrough and lived to be seventy-five, he was able to watch the changes that his invention made to the world. In fact, he grew tired of the telephone and after 1879 had no more to do with it, leaving improvements to others. He spent the rest of his life on inventions in other fields – none as successful as the telephone – and in working for the deaf.

FASCINATING FACTS

The first telephone subscriber in the world was an electrician in Boston called Charlie Williams. In April 1877 he had a line installed to connect his shop with his home.

The inventor of the automatic telephone exchange was a Kansas undertaker, Alman B. Strowger. He invented it in 1889 because he thought that operators were putting calls through to rival undertakers. The first system was installed in La Poste, Indiana, in 1892.

In 1877 there was just 2,600 telephones in the United States. Three years later there were 48,000, and ten years later over 150,000.

In 1892 Bell made the first long distance telephone call from New York. He spoke to Chicago, Illinois – a distance of about 1,000 miles.

The first telex machines appeared in 1932 and modern fax machines arrived in the early 1980s.

The telephone was much appreciated in North America where cities were much further apart than in Europe. The wires stretching across the prairies brought civilization nearer.

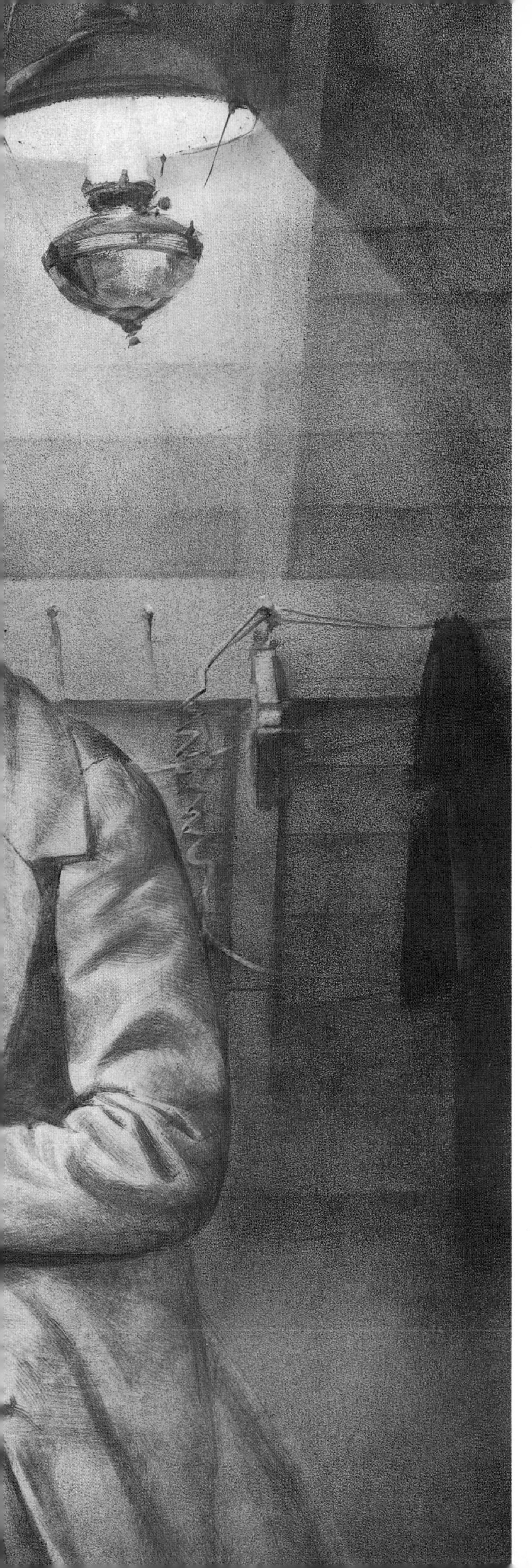

Tom Watson also left the telephone behind, becoming a successful shipbuilder. But they saw their "baby" grow from strength to strength.

Wires around the world

The first telephone exchange – it had twenty-one names, which were printed in a "directory" on one sheet of paper – opened at New Haven, Connecticut, in 1878. In 1892 the first automatic exchange, with dial telephones, opened. Gradually, the network of long-distance telephone wires spread across North America, although it was not until 1915 that New York, in the east, could speak direct to San Francisco in the west.

By the time Bell died in 1922, there were about 28,000,000 telephones in use in the world. He lived long enough to see the invention of the radio – the start of the next communications revolution. When he died, plans were being made to join telephone and radio in the first voice communication between North America and Europe.

However, it was not until September 1956 that a transatlantic telephone cable came into operation. Direct dialling from Europe to North America started in 1971. Since then, telecommunication satellites (see pages 82–85) have made most areas of the world reachable by telephone and have improved the quality of what you hear.

Alexander Graham Bell spent as much time as he could spare from his job as a teacher of the deaf working in the two rooms in downtown Boston that served as his "laboratory." There, he stumbled by accident on the way of turning sound vibrations into variations of electric current, which was the basis of the telephone.

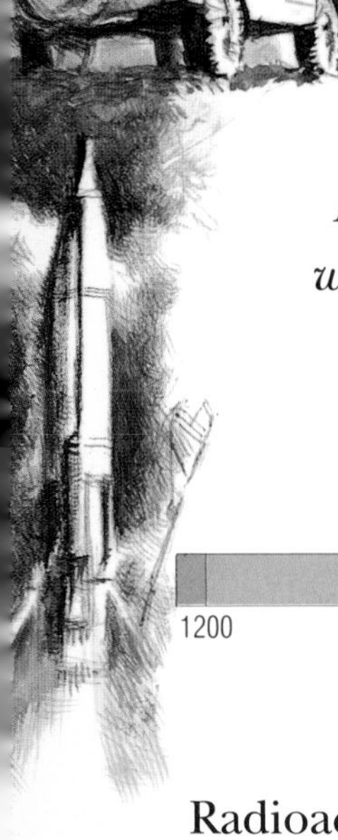

Marie Curie & Radium

More than three years of painstaking work had paid off for Marie Curie when she isolated a sample of an unknown element, radium, which had a strange, almost magical, power – she named it radioactivity. From her pioneering research came X rays, radiotherapy for cancer, atomic power stations and a deeper understanding of how the universe began.

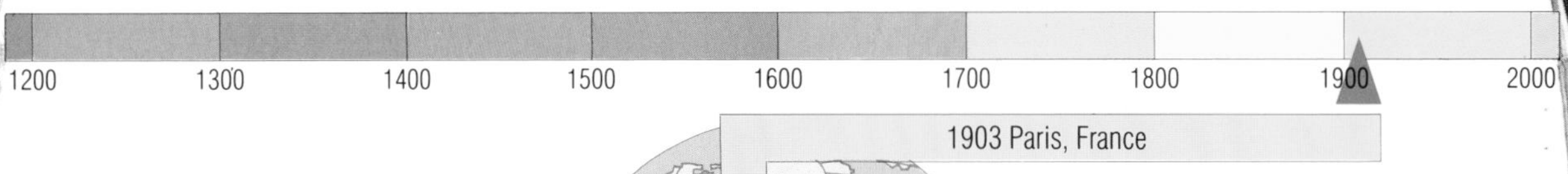

1200 1300 1400 1500 1600 1700 1800 1900 2000

1903 Paris, France

Radioactivity and radiation are familiar words today. Everyone knows about the dangers of radioactive leaks from nuclear power stations, and about the benefits of radioactive treatment for cancer and other diseases. Yet one hundred years ago the word was unknown. It was chosen by Marie Curie (1867–1934) and her husband Pierre to describe the strong electrical currents given out by certain chemical elements.

Polish pioneer
Marie Sklodowska was a pioneer in more ways than one. The daughter of a Polish teacher of mathematics and physics, she left Poland in 1891 to study physics at the Sorbonne in Paris. In the 1890s, it was rare for a woman to go to university at all. For a woman to study science was almost unknown.

In Paris, Marie met Pierre Curie, a lecturer in physics. They married in 1895, and from then on they worked together. Unusually for the time, although Pierre was the senior in years and position, it was Marie who led the partnership and, against the ingrained prejudices of the masculine scientific world, gained the greater reputation.

Marie Curie used her precious radium to power X-ray units like this one for the French army during World War I.

"It shines in the dark!"
Marie Curie's field of study was inspired by reports of the recent discoveries of X rays and of the radiation given off

Marie Curie's work on radium was a textbook example of scientific method. She had little equipment and few resources, but made up for this with tireless attention to detail. The support of her husband Pierre was also vital to her eventual success.

Pierre Curie (1859–1906) invented this electrometer which measures electrical currents, no matter how tiny.

FASCINATING FACTS

The Curies processed over 1,000 pounds of pitchblende to produce less than .0035 ounces of pure radium.

❑

It was not until 1904, when she was already internationally famous, that Marie received any pay for her work or had a proper laboratory in which to do her research.

❑

Marie Curie was:
the first woman doctor of science in Europe,
the first woman to win a Nobel Prize,
the first person to receive a Nobel Prize twice,
the first woman to lecture at the Sorbonne, and
the first woman to be elected to the French Academy of Medicine.

❑

In 1935, Marie Curie's elder daughter, Irène, also won a Nobel Prize for Physics with her husband, Frédéric Joliot-Curie, for developing the first artificial radioactive element.

by uranium. Marie began testing other elements and natural materials for "radioactivity" in 1897. She first used the word in her notebooks at this time.

She found that a mineral called pitchblende was far more radioactive than it should have been from the amount of uranium it contained. This led her, in 1898, to identify two new elements, radium and polonium, in pitchblende ore. Of these, radium – a million times more radioactive than uranium – was the more important.

The problem was to obtain enough radium for further experiments. The Curies rented an abandoned wooden shed from the School of Physics where they processed tons of pitchblende waste to extract the pure radium. In dreadful and dangerous conditions, the work went on until, by 1902, Marie had a tiny tube of pure radium.

"A great scientist"

Marie's research had been for her doctorate, which she was awarded in 1903, but already her reputation was international. Marie continued with her research into the properties of radium after Pierre's death in 1906, but others were already developing the possibilities that the discovery of radium had opened up. Marie Curie eventually died of radiation sickness; she had been exposed nonstop for over thirty years.

By 1903, French doctors were experimenting with the use of radium to destroy cancer cells. This was the beginning of radiotherapy. Meanwhile, in Canada, Ernest Rutherford was starting the work that would eventually lead to nuclear power and the atom bomb.

The First Powered Flight

When Flyer 1 *took off from the beach at Kitty Hawk, few people could have realized what would be the result of that 118-foot (36-meter) hop. In 1903, it took three days to sail from Europe to America; now Concorde flies it in under four hours. In 1903 a holiday in Europe would have been unthinkable for most people; in 1992 more than 20 million people took vacations abroad.*

1200 1300 1400 1500 1600 1700 1800 1900 2000

1903 Kitty Hawk, N. Carolina, USA

On a windy December morning in 1903 a little group of people huddled together on the sand dunes outside Kitty Hawk in North Carolina. They were watching a strange contraption like an oversized kite. But there was something different about this kite. It had an engine – and there was a man on board, lying on his stomach between the wings.

Airborne

The contraption moved forward, lifted its tail, and left the ground. Twelve seconds later and 118 feet (36 meters) along the sand, it came to earth again. Orville Wright had become the first person to make a powered flight in a heavier-than-air machine. The age of flight had begun.

The machine, which they called the *Flyer*, had been built by Orville Wright and his elder brother Wilbur. The flight on that December day was the end of a challenge that had taken up all their spare time for five years. It was also the end of a challenge that had eluded inventors for thousands of years – the challenge of flying like a bird.

The Wright brothers were Americans: Wilbur (left) (1867–1912) was the "brains"; Orville (1871–1948) had the practical skills.

Dreams of flight

There had been many attempts at flight. People had strapped huge wings on to their backs and thrown themselves from cliffs, flapping like birds – with disastrous results. In Italy, in the sixteenth century, the great artist Leonardo da Vinci (1452–1519) had drawn designs for a helicopter and a machine like an aircraft, but they had never been built.

The first device to carry people into the sky was a hot air balloon which rose over Paris in 1783.

Attention later turned to gliders based on kite designs, and many inventors made successful unpowered flights. Two Englishmen even tried to build an aircraft powered by steam, but this was far too heavy to get off the ground. An American inventor successfully launched a model aircraft powered by steam, but without a passenger.

Fresh thoughts

The invention of the much lighter gas engine set engineers thinking afresh. In Dayton, Ohio, the Wright brothers, who ran a bicycle shop, became fascinated by the thought of flying. They pored over all the books they could find on the subject, and experimented with kites and gliders before they began building the gas-engined *Flyer.*

Although the first flight was successful, there was a setback after the landing. A gust of wind caught the *Flyer* and wrecked it. Undeterred, the brothers set to and built another *Flyer,* and then another. By 1906 they had an aircraft that could stay in the air for thirty-eight minutes and cover nearly 25 miles (40 kilometers). That year, they set up an aircraft-building business.

Aircraft at war

News of the Wright brothers' exploits sparked off a "flying craze," especially in Europe. The military were among the first to see practical uses for aircraft. World War I, which broke out

One of the great moments in history, as Orville Wright takes his first twelve-second flight. The history of aviation began here, and leads in an unbroken line to the airliners and supersonic combat aircraft of today.

Charles Lindbergh (1902–74) was born in Detroit, Michigan. In May 1927 he flew solo nonstop in this Ryan monoplane, called Spirit of St Louis, *from Long Island in New York to Paris. It took him thirty-three and a half hours, and was almost twice the distance flown by Alcock and Brown in their nonstop flight in 1919.*

in 1914, was the first war in which aircraft took part. At first, the crews merely observed enemy positions and movements, but before long aircraft were being used to bomb and machine-gun troops and civilians, and for aerial combat.

The war brought about rapid developments in aviation, and when it was over there was a surplus of aircraft, pilots and skilled engineers. Some of these planes were used to set up the first airmail and airline services, or to popularize flying by taking people on short "joy-ride" flights. The pilots made record-breaking flights, which were keenly followed by the press.

The trailblazers

The years between 1918 and 1939 were a period of great excitement in the aviation world. In 1919 a U.S. Navy flying boat made the first crossing of the Atlantic, stopping on the way at the Azores. In the same year, two British aviators, John Alcock and Arthur Whitten Brown, made the first nonstop crossing. In 1927, an American, Charles Lindbergh, was the first to fly solo across the Atlantic.

Steadily, through the 1920s and 1930s, trailblazing flights by both men and women aviators broke record after record. An American, Amelia Earhart (1897–1937), was the first woman to fly the Atlantic, in 1928. In 1932, she was the first woman to fly it solo, and also the first person to fly it twice.

Amy Johnson (1903–41) was a British pilot who made the first solo flight by a woman from England to Australia in 1930. In 1932 she made a record-breaking solo flight to Cape Town in South Africa and back. She was also the first woman to gain a certificate as a ground engineer.

The age of air travel

Just as World War I had acted as a spur to progress in aviation, so did World War II. Thousands of large metal-framed bombers were built, and by 1944, propeller engines began to be replaced by jets. When peace came in 1945, the scene was set for the growth of regular air services crisscrossing the world and special charter flights taking parties of people on vacation abroad.

By the 1960s, flights between continents spelled the end of passenger shipping services across the oceans, and internal flights within countries dealt a serious blow to railway travel, especially in the U.S.

The introduction of the supersonic aircraft *Concorde* (which travels faster than the speed of sound) in 1976 meant that it was possible to land in Europe at an earlier time than the aircraft had taken off from New York. (This is because *Concorde* takes about three hours thirty minutes to fly across the Atlantic and the time zones are four hours apart.)

Seeing it all

Wilbur Wright died suddenly in 1912, but Orville lived on until 1948. So the first human being to take a powered flight saw aviation grow from a hand-built plane on a salt-flat in North Carolina to a worldwide industry .

Aircraft had become part of the terrifying weaponry of war. But they were also giving millions the chance to travel across the world on business or pleasure. They were taking injured people to hospital, delivering urgently needed drugs and supplies and helping farmers with their crops. Orville Wright lived to see the airplane conquer the world.

A Montgolfier hot-air balloon.

FASCINATING FACTS

The first men to fly were the Marquis d'Arlandes and Pilâtre de Rozier, who made an ascent in a hot-air balloon on November 21, 1783. The first woman to fly in a balloon was Jeanne Labrosse-Garnerin, in 1798.

❑

The first helicopter flight was made at Lisieux, France, by Paul Cornu on November 13, 1907.

❑

Lieutenant Thomas Selfridge of the U.S. Army was the first person to die in a plane crash, in 1908. He was killed when a plane piloted by Orville Wright went out of control; Orville was badly hurt.

❑

The first woman to fly in an airplane is reported to be a Madame Berg, the wife of a French businessman. She met Wilbur Wright when he visited France in 1909.

❑

A Frenchman named Adolphe Pégoud (1889–1915) invented the sport of aerobatics in 1913. He was the first man to fly an aircraft upside-down, and to loop-the-loop.

❑

The first aerial combat took place on October 5, 1914, between a French Aviatik and a German Voisin. The Voisin crashed and its two-man crew were killed. The first combat flyers were armed with revolvers and rifles.

Henry Ford & His Model T

Imagine how it must have been when you had to walk three miles (five kilometers) to school and back every day, when a shopping trip to the nearest town was a major expedition. When Ford built his first "Tin Lizzie," he gave millions of people a cheap and reliable car, which meant that a day's journey became hundreds, not tens of miles.

1200 1300 1400 1500 1600 1700 1800 1900 2000

1913 Detroit, Michigan, USA

Henry Ford was a young man in his twenties in 1885 when Carl Benz demonstrated the first car driven by a gas engine in Germany. It was a three-wheeler – slow, uncomfortable and difficult to drive. No one could have seen this as the start of a major new industry which was to change the world's way of life.

The thrill of something new

As Benz demonstrated his car, Ford was working in Detroit, Michigan, as a steam engine mechanic. The news from Europe about Benz's car, and of rival vehicles quickly developed by other engineers, excited Ford. Before long he had started to build a gas engine; then, in 1896, he built a complete car, of his own. It ran on four bicycle wheels – he called it the Quadricycle.

Then, in 1903, after having gained some experience working for another motor company in Detroit, Henry Ford set up his own business, the Ford Motor Company.

Henry Ford (1863–1947) pioneered production-line techniques. He turned car-making into a major world industry.

A car for everyone

Ford was much more than a brilliant engineer. He was also a clever businessman. When he looked at the growing car industry, he saw that most manufacturers were aiming at the rich motorist who wanted an expensive and dashing model for pleasure motoring. Ford looked beyond this to the time when every family would want a car for everyday use – a relatively inexpensive car that was easy to drive, cheap to run and easy to repair. He aimed to produce a car for ordinary people – what is now called the "mass market."

Other American car-makers, such as Russell Olds who produced a popular car called the Merry Oldsmobile, were also aiming at the mass market, but Ford singled out a special slice of the market – the millions of small American family farmers like his own father.

A reliable workhorse

These were people who needed a mechanical workhorse that could do a variety of jobs: take a load of cattle feed along rough unmade tracks, carry grain to market, go to the nearest town to pick up mail and shopping, or take the family to a party at a nearby farm.

Simple farmers did not want a fast or glamorous car. They wanted one that was reliable, and basic enough to maintain and repair themselves. Above all, they wanted a car that they could afford. Henry Ford made it his aim to meet their needs.

The Ford Motor Company's first car was the Model A, which appeared in 1903. Other designs followed, but the big breakthrough was still to come, though Ford was already working on it.

In 1906 his company moved to a new factory, which would be used to make the "breakthrough" car which Henry was designing. On October 1, 1908 the new model went into production. It was the Model T Ford.

On a winner

The first Model Ts cost $850. This was not cheap, but the Model T was stronger, cheaper to run, easier to maintain and more generally useful than other cars at that price.

Orders for cars poured in faster than the factory could produce them. Henry Ford had a winner on his hands but lacked the ability to satisfy the demand. In an attempt to meet the orders, he announced in 1909 that his company would in future make only

The Ford factory at Highland Park opened in 1906.

FASCINATING FACTS

Over 1,000,000 Model T's were manufactured in 1922 – the first year that this had happened.

The final total of Model T's manufactured was 15,007,033. In addition, millions of spare parts were made to keep the Model T's on the road.

This total of 15,007,033 cars was the world record for the production of a single model, until in 1972 it was broken by the Volkswagen "Beetle."

Ford's production methods were copied in Europe after Giovanni Agnelli, of the Italian company Fiat, visited Highland Park in 1912. Soon after his return to Italy, plans were made for a new Fiat factory near Turin.

Model T's, and that the customer could "have a car painted any color that he wants, so long as it is black."

Despite this change, the factory still struggled to keep up with the flow of orders. It hurt Ford to think of the business he might be losing because he could not satisfy the demand. He realized that people who could not buy a Model T would go away and buy a car from another company. By now, there were other manufacturers, large and small, who were trying to cash in on the same market. Ford also saw that if he could speed up production and bring the Model T's price down, the market would be even larger.

Mass produced parts

To solve these problems, Henry Ford turned his attention to the style of manufacturing which is called mass production. The old way of making mechanical products, such as clocks

and watches, was to build each one by hand. Craftsmen would shape each part separately, so that each one had slight individual differences.

Mass production means assembling products from identical parts which have been made beforehand and which can be interchanged. This not only makes production faster but also means that repairs can be made because spare parts are available.

Mass production was not new. The

This picture shows an early stage in the development of mass production at the Ford factory. The process of making a car has been broken down into single steps which are arranged in order for ease and speed of working. The car body, with the back seat already fitted, is slid down the ramp as the chassis rolls underneath. Fitters stand by to bolt the body in place and then the car will move on to have finishing touches added.

A 1913 Model T Ford, made on the new assembly line.

idea had been used since 1800 in the manufacture of guns, and later to make clocks, watches, sewing-machines and typewriters. Russell Olds had adopted it in 1902 to make his Merry Oldsmobiles on an assembly line. But Henry Ford added something new.

Ford's assembly line was a moving conveyor belt along which the car travelled as each new part was added. Each worker had a small, specialized job to do as the car passed by. Each new part was brought to the worker by a smaller conveyor belt. Parts and half-finished cars no longer had to be trundled around the factory floor. Production leapt dramatically after 1913, when the conveyor belts were introduced, but another crisis lay just around the corner.

Trouble on the line

Ford's workers did not like the assembly line. They were making more cars, but earning no more money. Soon they began to leave for other less boring jobs. Production began to slow down because there were not enough workers to run the assembly line.

So Ford set up a scheme for workers to share in the profits of the company. The more cars they made, the more money they would take home. The scheme doubled the workers' pay. Once again, Ford had found an answer to an industrial problem.

From then on, the Model T went from strength to strength. By 1916, 2,000 cars were being produced each day, and the price had fallen to around $360. In 1922 Ford broke through the "million barrier" and over 1,200,000 Model T's were made. It was not until 1927, with over 15,000,000 Model T's on the roads, that production ended and new models were introduced. But well into the 1950s, American small farmers were still using their Model T's, sturdy and dependable to the end.

Einstein's Theory of Relativity

Who would have expected an official in a Swiss post office to change the whole basis of scientific thought? But one did! Einstein's theories of relativity forced a radical rethinking of how the universe works and is put together, and opened new fields of scientific knowledge. And he worked all these ideas out in his head and on scraps of paper!

1200 1300 1400 1500 1600 1700 1800 1900 2000

1915 Berne, Switzerland

Most important scientific discoveries have been made by people carrying out experiments and then repeating them over and over again to see if they get the same results each time. This work is usually done in universities or in research laboratories of large companies.

Albert Einstein completely changed scientific thought without carrying out a single practical experiment. He worked on his own with pencil and paper, and did his experiments in his head.

Travelling light

The first results of Einstein's work was published in 1905. He called it the Special Theory of Relativity, and it was concerned with the velocity (speed) of light. Most forms of velocity are relative. This means that an object moves in relation to something else. Imagine a train travelling towards a bridge at 40 miles per hour. Its velocity is 40 mph relative to the bridge. Now imagine that someone throws a ball at 10 mph from the train in the direction of the bridge. The ball will have a velocity of 50 mph when seen from the bridge – the speed of the train plus the speed of the throw.

Einstein's theory said that light does not behave in the same way. It always travels at the same velocity, so its speed does not change whether the source of light is moving towards you, away from you or standing still.

Albert Einstein (1879-1955) worked out most of his theories by doing "thought" experiments. These could only be done in his imagination.

The faster the smaller

From this starting-point, Einstein came up with some surprising ideas. He said that if an object's speed approached the speed of light, it would become smaller in size, but its mass (or weight) would increase. This change

would increase the object's energy, because a heavier object has more energy than a lighter one moving at the same velocity. No object can travel at, or faster than, the speed of light.

Another idea was even more surprising, because it seemed to go against common sense. Imagine that there are a pair of twins, one of whom takes a journey in space, travelling near the speed of light. This journey lasts seventy Earth years. Einstein's theory said that time would pass more slowly on the spacecraft and so only ten years would pass. The twin on the spacecraft would only age by ten years, while the twin left on Earth would age by seventy. This sounds very unlikely, but laboratory experiments have proved it to be true.

A deadly conclusion

The theory also included the world-famous equation which said that solid objects could be converted into pure energy: $\mathbf{E = mc^2}$. In this equation **E** is energy; **m** is mass; and **c** is the speed of light. One result of this discovery was nuclear power and the atom bomb.

Bending time and space

Ten years later, in 1915, Einstein published his General Theory of Relativity, which says that gravity should be thought of as the bending of space and time. This means that acceleration (an increase in speed)will have the same effect on an object as

Einstein was a theoretical scientist. He left it to others to carry out the experiments that would prove his theories correct. Many of these experiments had to wait until instruments had been developed which were accurate enough to measure the phenomena Einstein had described.

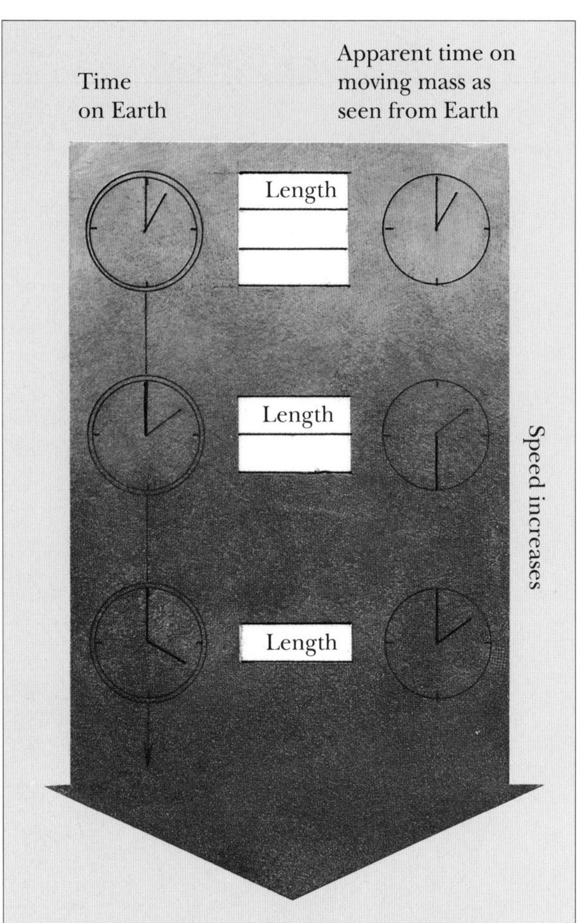

A clock on an object accelerating to very high speeds will run slower than one on Earth. As the object goes faster, the clock runs slower.

FASCINATING FACTS

If two aircraft take off at the same time and place, one going eastward and one westward, a clock in the eastward-bound plane will appear to run slightly slower than one in the other plane. This is because the Earth spins from west to east, and so the first clock travels further.

❑

If you drop a ball inside a lift that is falling at the speed of gravity, the ball will appear to "hang" in the air. This is because the lift and the ball are falling at the same speed.

❑

Einstein deduced that there are no straight lines in space, because of the force of gravity. This means that space travellers who set out on a straight line, will eventually return to their point of departure – they cannot "escape" from our universe.

gravity does. This was a major milestone in science, because it corrected Sir Isaac Newton's ideas about gravity (see pages 24–26).

Since Einstein's death, spaceship crews have proved this theory to be true. When a spaceship's rocket engine lifts it off the Earth, the crew experience the same feeling as when they pass by a planet and the pull of its gravity affects the spacecraft.

Einstein also corrected Newton's theories about the orbits of the planets round the Sun. Newton said that these orbits were elliptical, like a flattened circle, and that they were fixed. The planets moved in orbit, but the orbits themselves did not move. Einstein showed that the orbits themselves rotate slowly in space.

A pacifist at war

The splitting of the atom, what to expect when astronauts went into space, the make-up of the universe itself – all these ideas and more sprang from Albert Einstein's work. This covered the whole of physics, from the behaviour of the smallest particles of matter to the structure of the universe.

But his life was not all theory. In 1914 he was made Research Director at the Institute of Physics in Berlin, Germany, but he had to leave in 1933, after the Nazis came to power, because he was Jewish. He went to live in the United States.

In 1939, despite being a pacifist (someone who rejects war), he wrote to President Roosevelt warning him that it was now possible to build an atom bomb and that Nazi Germany was planning to do so. This started a race to build the first atom bomb – a race which the United States won.

The Inventors of Television

You will find televisions everywhere you go – from the Russian steppes to the Australian outback, from the Amazon jungle to a coral island. You can watch cricket from India, a UN debate or animals from an African game park in your own living room. How many of the shoppers who saw this novelty in London in 1923 guessed this would happen?

1200 1300 1400 1500 1600 1700 1800 1900 2000

1925 London, England

You would have laughed to see John Logie Baird's first television camera and receiver. He made them out of odds and ends. There was a lamp inside a biscuit tin and lenses from old cycle lamps. The parts were mounted on wood offcuts and held on with string. You would never have believed that such an outlandish piece of equipment could work.

The first TV picture

But it did. The first picture it produced showed the flickering shape of a cross. That was in 1924. Within a few months Baird had produced an improved model which transmitted the image of a puppet and then of a real person. Soon afterwards, Baird demonstrated television to an audience of distinguished scientists and then to shoppers at a London department store.

John Logie Baird (1888–1946) was born in Scotland. In 1925 his office boy became the first person ever to appear on television.

Rival systems

John Logie Baird was a Scottish engineer and businessman, but bad health forced him to retire in 1923 when he was only 35. Soon afterward, he began his experiments in transmitting pictures.

The method that he used to make television pictures was entirely different from the system used in our television sets today. This was invented by an American, Vladimir Zworykin (1889–1982). Modern sets have a cathode-ray tube, which scans a chosen scene electronically and then produces a picture made up of hundreds of lines.

Baird used a metal disk with a spiral of holes cut in it to scan the picture mechanically. The light passed through the holes to photoelectric cells which changed it into electrical signals. A second disk turned the signals back into pictures.

Baird's "Televisor" sets were substantial pieces of furniture, showing their status as luxury items.

FASCINATING FACTS

The first television sets had tiny screens not much larger than a postcard. This was because of the technical difficulties of making large cathode-ray tubes. You had to sit very close to see what was being shown.

The first electronic television camera and receiver were invented in the 1920s by Vladimir Zworykin, a Russian working in the U.S. It was his system that was adopted instead of Baird's.

Today, television pictures are bounced around the world by satellite. But in 1927 Baird sent pictures from England to the U.S. using earth radio stations.

Television pictures were transmitted in black and white only until the 1960s. However, in 1928, Baird demonstrated a color TV system, which merged red, blue and green images to make a full-color picture.

The result was a crude, flickering picture made up of thirty lines.

Other inventors were working on electronic systems, though Zworykin's cathode-ray model for RCA did not appear until 1932. Baird was the first to produce a recognizable picture, and he persuaded the British Broadcasting Corporation (BBC) to start regular broadcasts in 1929 using his system.

About 10,000 people bought his "Televisor" sets to receive the programs. But disappointment lay ahead. Development of the cathode-ray system in America was catching up fast. The pictures were much better than Baird's, and soon the BBC started transmitting on the cathode-ray system as well. Finally, in 1937, the BBC dropped Baird's system altogether.

Hurt pride

This was a tremendous blow to Baird. It was not just that he would have made a fortune if his system had been adopted. The BBC's decision proved that he was wrong in sticking stubbornly to mechanical scanning. He later worked with cathode-ray tubes and his results were equally successful.

Despite this, it was Baird's pictures that began the television age. His flair for publicity created the excitement that spurred engineers on to produce a better system. Baird was the true "father of television."

Baird's early television experiments, like this one using a puppet's head, produced only a blurred black-and-white image on a tiny screen. It was impossible to watch the screen for long.

Building the First Computers

"If you can't decode these messages, we shall lose the war!" That was the task that Churchill set Alan Turing's team in 1941. From that challenge the first computers were born, the ancestors of the millions of computers that run the world today. Every time you play a Nintendo game, or call Directory Assistance, or shop in a bar-coded supermarket, think of Alan Turing who made it possible.

1200 1300 1400 1500 1600 1700 1800 1900 2000

1943 Bletchley, Bucks, England

The idea of using a machine to make your calculations is many thousands of years old. The Chinese invented the abacus or counting-frame as long ago as 800 B.C.

In 1832, an Englishman called Charles Babbage (1791–1871) conceived his Analytical Engine, but it was never built in a workable form. It is usually considered to be the first computer design.

By 1900, there were a number of calculating machines for use in offices, but these could only add, subtract, divide and multiply. They could print out their calculations, but they could not store information or process it.

Dawn of a revolution

In 1937, an American, Howard Aiken (1900–73) constructed the IBM Mk7 – a highly advanced electronic calculator. In the same year, a young British mathematician, Alan Turing (1912–54), published a theoretical description of a computer system that could store information and process it, using punched paper tape. The paper, "On Computable Numbers," changed Turing's life. From then on, he was involved with the development of computers. His paper started the revolution that has made the computer a part of everyday life.

The "Enigma" encoding machine. The first computers were developed to let the Allies break the Nazis' codes and overhear their battle plans.

Unlocking secrets

The first real computer to process information according to a program was built in 1943 with Turing's help. It was produced in response to

Alan Turing stands in front of Colossus 1, the computer which he helped to build. The punched paper tape he is holding is the computer's program. Long rolls of it had to be fed through Colossus before any calculations could be made.

Charles Babbage designed his "Analytical Engine" to make decisions according to the results of its calculations.

FASCINATING FACTS

The first all-purpose, stored-program computer was called ENIAC. It was built in 1945 in Philadelphia, Massachusetts. It contained 18,000 valves and weighed about 33 tons (30 tonnes).

❏

A light pen allows a computer user to draw pictures and designs on the screen. This is called computer graphics, and it was invented in 1960 by an American electronics engineer, Ivan Sutherland.

❏

The first voice-operated computer that could reply in a synthetic voice was built by EMI-Threshold in Great Britain in 1973. The idea was invented by two Americans, Thomas B. Martins and R.B. Cox, in 1973.

❏

Some integrated circuits contain more than 1,000,000 components on a silicon chip about .2 inches (5 mm) square. Each chip is tested at the factory. If any one of the components is faulty, the chip is scrapped.

❏

A computer fitted with a device called a modem can communicate with other computers thousands of miles away through the telephone network.

wartime needs. During World War II, the Allies needed a means of decoding enemy military signals – in particular those produced by a Nazi code machine known as "Enigma."

Working in secret, a team of crack British scientists built a computer, which they named *Colossus I*, at Bletchley in Buckinghamshire. At about the same time, the U.S. Navy paid for the building of an American computer, *Harvard Mark I*, which was also used for military purposes.

Faster and smaller

These machines were huge. Their electronic circuits were made up of valves connected by miles of wires. These valves became very hot when the computers were running, and special cooling systems were needed. Compared with today's computers, these early models worked slowly, but they could make complex calculations far more quickly than human beings.

Since then, the computer industry has made computers faster, smaller, and able to perform a wider range of tasks. Transistors, invented in 1947 by three Americans, took up less space than valves and did not produce so much heat. Then came the integrated circuit, which combined large numbers of transistors and other components on one tiny silicon chip.

Millions of chips

These chips have made it possible to produce computers small enough to fit in the user's lap. Silicon chips are mass produced, so they are cheap enough to be used in personal computers and to program equipment in the home, such as video recorders, washing machines and compact disc players.

The Atom Bomb

"Look at the dust cloud," cried an observer in the B29, "It looks like a...a mushroom!" That cloud, rising from the devastated city of Hiroshima, sent a shiver round the world even as it rejoiced at the ending of World War II. Humanity had found a new power source – one that could bring both great benefits and enormous destruction.

1200 1300 1400 1500 1600 1700 1800 1900 2000

1945 Los Alamos, New Mexico, USA

The atom bomb that was dropped on Hiroshima in Japan on August 6, 1945, followed by another on Nagasaki three days later, marked the end of a deadly race. The bomb was the work of a team of scientists in America. Until Nazi Germany had surrendered in May 1945, her scientists had been working on a similar project. If they had won the race, the war might have ended differently.

Inside the atom

Ever since Marie Curie's pioneering work on radiation, physicists had been looking for the secrets of the atom. Ernest Rutherford made many of these discoveries including three different forms of radiation and the structure of the atom.

In 1917, he was the first person to split the nucleus that lies at the heart of an atom. This is known as nuclear fission.

Ernest Rutherford (1871–1937) was born in New Zealand. Called the father of nuclear physics, he discovered many secrets of the atom.

All through the 1920s and 1930s, physicists around the world found out more about atoms and their behavior.

The grim race

Then early in 1938 in Berlin, Lise Meitner (1878–1968) and Otto Hahn (1879–1968) succeeded in splitting the nucleus of a uranium atom. What was more, they theorized that, once started, the process would continue in what was called a "chain reaction", releasing huge amounts of explosive energy. Meitner fled from the Nazis to Sweden later that year. There she continued her work on atomic energy.

By 1939, German and American physicists were trying to turn the theory of nuclear fission into practice. Warned by Einstein, the American Manhattan Project, led by Robert Oppenheimer (1904–67), aimed to develop an atom bomb.

The mushroom cloud of dust raised by the energy unleashed by the first atom bomb towers over Hiroshima on August 6, 1945. The danger of nuclear weapons does not lie only in their explosive force. They release radiation into the atmosphere – this can cause cancer and genetic damage causing babies to be born with defects.

Hiroshima was the result. The bomb, dropped from a U.S. Air Force B29, destroyed most of the city and killed over 200,000 people. Japan surrendered and World War II was over in the Far East as well as in Europe. But the world still trembled.

In 1945, only the U.S. had mastered the technology of nuclear weapons, but other countries wanted it too. First the USSR (as it was then), next Britain, and later France, China and India developed atomic weapons.

No safety on Earth

The USSR and the U.S. invested the greatest resources in nuclear weapons. These two "super-powers" constantly tried to extend their influence around the world. Both built up stocks of weapons large enough to blow the world to pieces. Any increase in tension between the super-powers could have led to World War III.

Meanwhile, even more destructive nuclear weapons were developed. In

FASCINATING FACTS

The first test explosion of an atom bomb took place at Los Alamos in New Mexico, in July 1945.

❑

The nearest the world came to World War III was in October 1962. The USSR was building missile launch sites in Cuba, close to the coast of the U.S. U.S. Navy ships lined up to stop Soviet ships delivering the missiles. At the last moment the Soviet ships turned back.

❑

By 1988 the U.S. had over 14,000 nuclear warheads, which could be delivered by rockets or aircraft. The USSR had over 11,000. Both countries also had thousands of smaller nuclear weapons.

There are about 400 nuclear power stations around the world. Some of their waste products will remain radioactive for more than 13,500 million years.

❑

The radioactive element Carbon-14 is found in all living things – plants, animals and humans. By measuring how much Carbon-14 remains in a fossil, bone or object, scientists can work out how old it is.

❑

The worst nuclear accident happened on April 26, 1986 at Chernobyl in the Ukraine (then in the USSR). A nuclear reactor exploded, killing at least 19 people and contaminating a wide area of the Ukraine. The radioactive fallout reached Britain.

How nuclear fission works: a slow-moving neutron (yellow) is launched so as to split a uranium atom (pink). Each split uranium atom gives off three more neutrons,plus two hydrogen atoms (red). The new neutrons will then split other uranium atoms – this happens over and over forming what is called a "chain reaction."

1952 the U.S. exploded the first hydrogen bomb; the USSR followed in 1953. These work by forcing together (or fusing) hydrogen atoms, which then give off explosive energy.

Real "Star Wars"

During the 1950s, both powers perfected rockets as a way of delivering nuclear bombs. Then came the "space race," and the terrible possibility that weapons might be stored in space and fired at any time by signals from Earth.

It seemed to ordinary people that the world's leaders were determined to destroy it. Anyone who survived a nuclear war would almost certainly die later from the radiation. In western Europe and North America, thousands of people took part in demonstrations calling for the abolition of nuclear weapons. But the weapons existed – they could not be "uninvented."

An anti-nuclear demonstration in the 1960s.

Controlling the monster

The only solution was to control the number and use of nuclear weapons. In the late 1980s, after long years of negotiations, the super-powers started reducing the number of weapons they held. The break-up of the USSR in 1991 has reduced the risk of a nuclear war, but the threat is not over.

Two worries remain. One is that a weapon might be set off by accident and unleash a nuclear war. The other is that a state with nuclear power, but no bombs, might develop them so as to blackmail the rest of the world.

Power from the atom

However, the work of nuclear scientists has not all been about developing weapons. The controlled release of energy in a nuclear reactor can also be used to generate electricity. In 1942, a team of physicists in Chicago, Illinois, led by an Italian physicist, Enrico Fermi (1901–54), achieved the first controlled chain reaction in a reactor.

By 1943, the first commercial nuclear reactor was making electricity at Oak Ridge, Tennessee. Since then nearly 400 nuclear power stations have been built. Both France and Sweden produce over 40 percent of their electricity by nuclear power.

As the supplies of fossil fuels (oil, gas and coal) begin to dwindle, more nuclear power stations will have to be built.

The Decoders of DNA

Why does a person have blue eyes or tightly curled hair or olive skin? The answer lies in the genes they inherit from their parents, which tell the baby how to develop. The information is encoded in millions of DNA molecules, found in every cell of your body. Scientists knew if they could read this code, they could learn much more about how our bodies grow and work.

1200 1300 1400 1500 1600 1700 1800 1900 2000

1953 Cambridge & London, England

When plants or animals reproduce, cells from a male and female come together to form a new living organism. Each cell contains thin threads, called chromosomes. These are made up of molecules of deoxyribonucleic acid, or DNA. Part of these chains are the genes.

Genes carry coded information about the characteristics of the new living creature's parents. This information determines what the new creature will look like. For example, if your eyes are like your father's or your nose is the same shape as your mother's, it is the information in your genes that have made them like that. This passing on of features from parent to child is called heredity.

In a monastery garden
The way heredity works was discovered by a monk, Gregor Mendel, after he had studied generations of pea plants in his garden. Later, the Danish biologist Wilhelm Johannsen (1857–1927) found that chromosomes and the genes inside them were responsible for heredity. But what were the genes made of, and how did they work?

Gregor Mendel (1822–84) was an Austrian monk who discovered the rules of heredity through experiments in his garden.

Breakthrough
The beginning of the answer came in 1944 when scientists discovered that the DNA contained in the chromosomes carried the genetic message. But they still had to find out how the DNA molecule was put together.

In the early 1950s, two teams of scientists began to inch towards the solution. Maurice Wilkins and Rosalind Franklin in London were studying X rays of DNA, while in Cambridge, Francis Crick and James D. Watson were building a model of the DNA molecule. It had been established that

DNA contained a sugar, phosphate, hydrogen and four substances known as "bases." The trouble was that they could not fit these items together in a chemically logical way.

Then Watson and Crick rebuilt their model with the hydrogen atoms and bases inside two strands of the sugar and phosphate. These two strands coiled around each other in a double helix (like a twisted rope), while inside the four bases, linked in pairs with the hydrogen atoms, held the two strands together.

A greatly enlarged view of a single cell, showing the male and female chromosomes pairing up.

They compared their model with Franklin and Wilkins' X rays and were convinced. The breakthrough had been made at last.

Making alterations

After the structure of DNA had been decoded, scientists started to ask more questions. Could genes be transferred from one organism to another? Could they be stitched together to create entirely new plants or animals or even humans? This was the start of genetic engineering, the science of altering genes so as to change the characteristics of an organism.

Genetic engineering has already been carried out on plants to produce varieties more resistant to disease, and it has produced insulin for the treatment of diabetes in humans.

Scientists have also developed a technique called "genetic finger printing." This can identify the genetic make-up of each individual person and is used in the study of disease and in the detection of criminals. But one vital question, which has not yet been answered, is how far science should go in genetic engineering. Have scientists the right to interfere with nature or, as some people would say, "play God"?

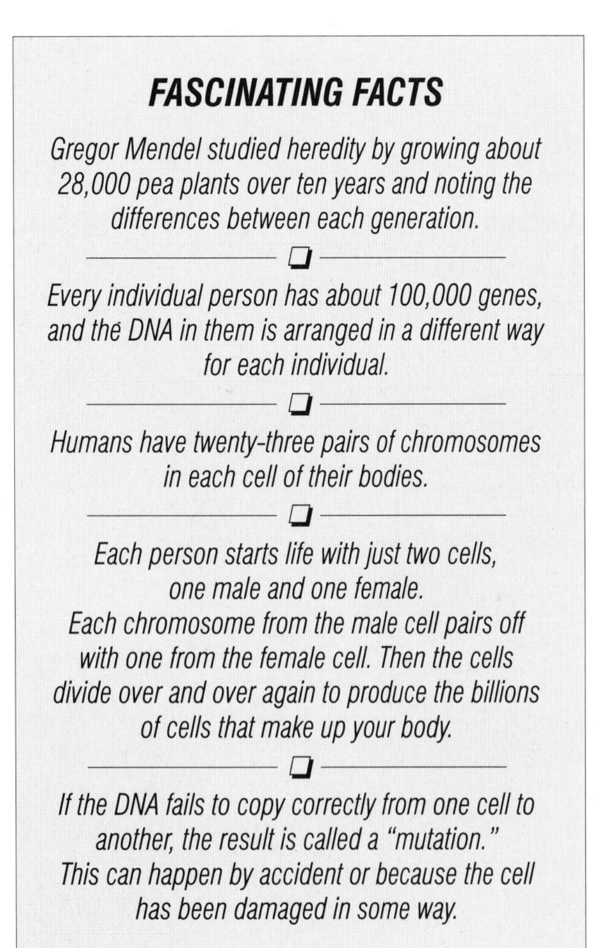

FASCINATING FACTS

Gregor Mendel studied heredity by growing about 28,000 pea plants over ten years and noting the differences between each generation.

Every individual person has about 100,000 genes, and the DNA in them is arranged in a different way for each individual.

Humans have twenty-three pairs of chromosomes in each cell of their bodies.

Each person starts life with just two cells, one male and one female. Each chromosome from the male cell pairs off with one from the female cell. Then the cells divide over and over again to produce the billions of cells that make up your body.

If the DNA fails to copy correctly from one cell to another, the result is called a "mutation." This can happen by accident or because the cell has been damaged in some way.

Nobel Prize winners James D. Watson, Francis Crick and Maurice Wilkins with the model of a DNA molecule. Millions of atoms are arranged in a double helix held together by cross-pieces. The order in which the atoms are arranged makes the "code" by which information is passed on. The missing member of the team, Rosalind Franklin, died at the early age of 37.

The Launch of Sputnik I

"Sir! Sir! Listen to this, Sir!" an excited British schoolboy shouted to his teacher. Those schoolboys had picked up a broadcast from the first man-made object in space. Since then hundreds of satellites have been launched into orbit with uses ranging from military via communications to forecasting the weather. They are essential to keeping the world running smoothly.

1200 1300 1400 1500 1600 1700 1800 1900 2000

1957 Tyuratam, Kazakhstan

The date was October 4, 1957. The main story on radio and television news programs all over the world was amazing. For the first time an object had been fired into space from Earth. To prove it, listeners could hear the radio bleeps it transmitted as it traveled round the Earth in orbit.

Secret launch

That bleeping object was *Sputnik I*, the first artificial satellite. It had been launched in great secrecy from a secret astrodome to the east of the Azal Sea near Tyuratam in Kazahkstan. This was another surprise, because most people had thought that the then USSR lagged far behind the western world in technology.

Sputnik I was tiny compared with today's satellites. It was about 22 inches (58 cm) across and weighed about 182 pounds. Its size was limited by the power of the rocket that put it into orbit. But within a few weeks the USSR launched *Sputnik II*, weighing half a ton. This carried the first living creature into space, a dog called Laika. *Sputnik II* was big enough to be seen in the sky without a telescope.

Rockets, powered by hydrogen or helium fuel, are used to break the bonds of gravity and lift satellites and the shuttle into orbit.

Race for space

Caught by surprise by the USSR's achievement, U.S. scientists worked desperately to catch up. Their first successful launch was *Explorer I* in February 1958. Then the "space race" was on.

The first man in space, Yuri Gagarin of the USSR, orbited the Earth once on April 12, 1961 in *Vostok I*. The U.S. replied with John Glenn, who made three orbits in February 1962.

The rivalry between the two space powers

continued nonstop as each advanced in different areas of space technology and then caught up with each other.

Science fiction come true
The launches that carried astronauts into orbit made the biggest headlines and caused the most excitement. It was like watching science fiction stories come true. But space flights with human beings on board were only one part of the space race. Another part had to do with the Cold War between the half of the world that looked to the USSR for leadership and the half that turned to the U.S.

The rockets and technology used to put satellites into orbit and guide them were the same as those designed to carry and aim nuclear weapons (see pages 75–9). The U.S. and the USSR were showing off their technology and giving a warning to each other. They also put into orbit reconnaissance or "spy" satellites to photograph each other's territory and transmit pictures back to Earth.

"Nation shall speak unto nation...."
There are many peaceful uses for satellites as well, and some of these have changed our lives dramatically. Perhaps the best-known are the communications satellites which beam television broadcasts into our homes. The radio waves that carry television signals on Earth work only over short distances because they cannot shape themselves to follow the curve of the Earth's surface. But signals beamed to a satellite and then transmitted again to receiving dishes can be "bounced" across the world. The first TV communications satellite was *Telstar I*, launched by the U.S. in 1962.

From top right anticlockwise: Landsat4 *(images of earth);* Voyager *(probe to Mars & Jupiter);* Intelsat *(telecommunications);* International Ultra Violet Explorer *(space telescope); the Shuttle;* Tracking & Data Relay Satellite; Sycom 2 *(first earth orbiter).*

FASCINATING FACTS

Among the first people to hear the bleeps transmitted by Sputnik I *was a group of students working in the laboratory of a school near Kettering in England.*

If satellites fall from their orbits, they re-enter the Earth's atmosphere and burn up. But the second U.S. satellite, Vanguard I, *launched in March 1958, is still in orbit.*

Pictures from satellites sometimes give scientists a surprise. One such surprise was that the Earth is slightly pear-shaped, and not as flat near the poles as they had thought.

The space race created a great deal of new technology, some of which now has everyday uses. One example is the chemical coating that makes saucepans "non-stick."

Other communications satellites bounce radio signals and telephone conversations around the world with far less interference than Earth radio or transoceanic cables. *Early Bird I* was the first commercial communications satellite, launched in 1965 by the U.S.

Satellites at work

There are now hundreds of satellites in orbit around the Earth. They carry instruments and equipment which provide a wide range of information. Some of them measure and report on conditions in space or take photographs of the Earth which are used in surveying and mapmaking.

Weather satellites report on world weather patterns, making forecasts more accurate. Navigational satellites send out signals which enable ships to plot their positions accurately. Astronomical satellites, like the Hubble telescope, send back clearer pictures of the universe than could ever be taken through the polluted atmosphere of the Earth.

Space colonies

Possibilities for the future are even more interesting. They may include the colonization of space by people who would live there permanently or for long periods in satellites. Space shuttles flying regular services like airliners would carry supplies and visitors to and fro. When *Sputnik I* went into orbit less than forty years ago, it began a new chapter in human history.

Moving at about 5 miles (8 kilometers) per second, Sputnik I *orbited the Earth in just over one and a half hours. The antennae seen projecting from it in this picture were its means of communication with the Earth.*

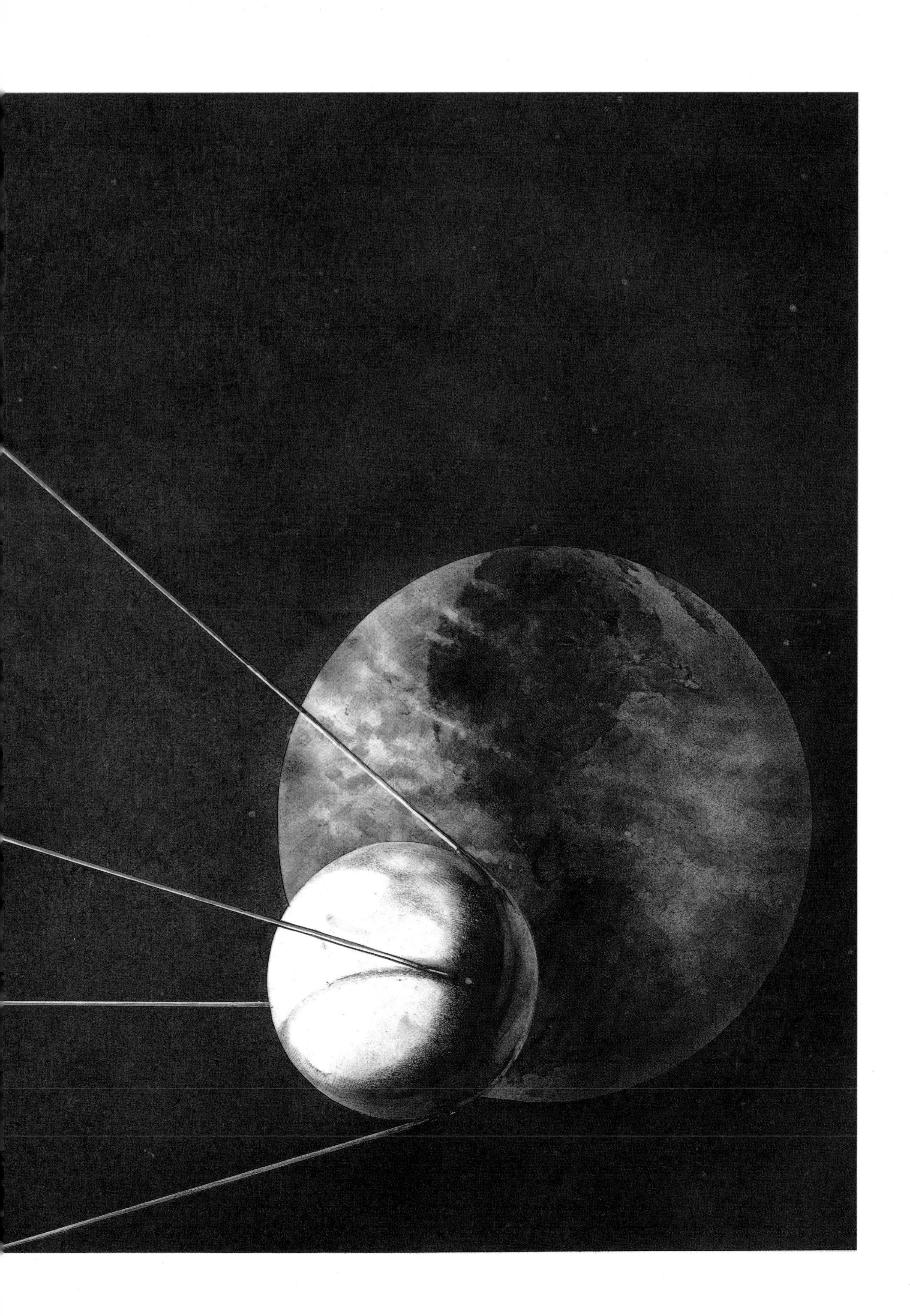

The First Men on the Moon

"One small step for man, one giant step for mankind." When Neil Armstrong climbed from the Apollo 11 *lunar module and became the first human being to stand on the surface of the Moon, the Earth was shining high in the sky like a moon. At that moment, people finally understood how tiny our Earth is compared to the enormous universe beyond.*

1200 1300 1400 1500 1600 1700 1800 1900 2000

1969 Cape Kennedy, Florida, USA

In 1961, the then President of the U.S., John F. Kennedy, made a promise which seemed unlikely to be fulfilled at the time. He announced the Apollo space program that planned to put a man on the moon by 1970.

By then, there had been voyages round the moon by unmanned Russian spacecraft which had sent back photographs of the far side, and *Lunik II* had crashed on the Moon. But no one knew whether the surface was hard or soft. The technology that was needed, involving a spacecraft that could be detached from the command module and then rejoin it, was untried. The Apollo program was a voyage into the unknown.

238,700 miles (384,500 kilometers) away

Many of the "unknowns" were answered as the Apollo program got under way. A series of U.S. probes in the mid-1960s crash-landed on the moon, but before they did so, they sent back close-up pictures of the surface. In 1966 two Soviet probes, *Luna XIII* and *Luna XIV* made soft landings and sent back more detailed information. Meanwhile, trial flights with Apollo command and lunar modules proved that the landing and take-off technology would work.

Finally, on July 20, 1969 – five months short of the deadline President Kennedy had set – Neil Armstrong and Buzz Aldrin stepped on to the moon's surface. Michael Collins stayed behind in the command module orbiting the moon.

Over the next three years there were five more landings by American astronauts. During the same period,

The surface of the moon was surveyed very carefully by unmanned satellites during the 1960s, before a manned landing was attempted.

unmanned Russian craft landed robot vehicles which explored over several miles.

Twelve brave men

Altogether, twelve astronauts – all Americans – set foot on the moon before the Apollo program came to an end in 1972. They needed enormous courage – compared with a Moon landing, an orbital space flight is almost as safe as an airline flight.

The critical moment for the moon travellers came when they took off in the lunar module to return to the command module. If the engine failed to achieve lift-off, there was not enough power for a second try. Any astronaut marooned on the moon would run out of oxygen long before a rescue mission could arrive from Earth. Fortunately, on each of the six moon landing missions the lunar module's take-off engine worked perfectly and all the astronauts returned safely.

New challenges

The information and samples the Apollo astronauts brought back gave scientists a very detailed picture of the moon and increased their thirst for knowledge of the planets. They were curious to know, for example, whether any form of life existed on them.

Venus is the nearest planet to Earth, which is why it shines so brightly in the night sky. It made good sense to start exploration of the planets there. But there is a big difference between a journey to the moon, only 238,700 miles (384,400 kilometers) away, and one to Venus, which at the nearest point in its orbit is 26 million miles (42 million kilometers) away.

FASCINATING FACTS

Pluto is the furthest planet from the sun but at times its orbit brings it inside the orbit of Neptune.

There are nine planets in our solar system. However, some scientists think that there may be at least one more, even further out than Pluto, which has not yet been spotted by astronomers.

Yuri Gagarin (1934–68), a Russian air force officer, was the first man to go into space in April 1961.

American space personnel are called astronauts; Russian space personnel and called cosmonauts.

The first woman in space was a Russian, Valentina Tereshkova. She flew in Vostok 6 *in June 1963. The first woman astronaut was Sally K. Ride, who flew in the shuttle* Challenger *in June 1983.*

The worst space travel accident was in January 1986, when the U.S. shuttle Challenger *blew up shortly after lift-off. Seven astronauts, including one woman, were killed.*

It took the Viking *probes ten months to reach Mars, while* Voyager 2 *took twelve years to reach Neptune.*

The first successful "flyby" Venus probe was the American *Mariner 2* in 1962. There was an unmanned landing by the Russian-built *Venera 8* in 1972.

Then, in 1975, two more Russian probes (*Venera 9* and *Venera 10*) landed on the surface and sent back pictures for about an hour before they stopped working. They were probably destroyed by acids in the rocky surface, or by the great heat. The most recent visit by an American orbiting satellite, *Magellan*, has produced a detailed map of the whole surface of Venus.

The red planet

Mars was the space explorers' next target. It is the second closest planet to Earth, next out from the sun, and is about 14 million miles (22 million kilometers) away at the nearest point of its orbit.

Mars has always been of special interest to astronomers because they thought that conditions there were most likely, among all the planets, to support some kind of life. However, they were only expecting to find very simple plant or animal organisms.

In 1877 an Italian astronomer called Giovanni Schiaparelli (1835–1910) saw what he thought were canals on the surface of Mars. If they really were canals, someone must have made them. This led to the idea that there was, or had been at some time, a civilization on Mars.

Aiming for Mars

The first "flyby" probe to Mars was launched in 1965, and was followed by others which sent back photographs of the surface. In 1971 a U.S. probe went into orbit round Mars, and there was a crash landing by another Russian

space probe in 1975. In 1976 the U.S. *Viking 1* and *2* made the first soft landings by unmanned space vehicles.

None of the instruments on either of the *Viking* landers could find any signs of life, even of microbes in the soil. This was not surprising, because most of the atmosphere of Mars is made up of carbon dioxide, and there is no liquid water on the surface.

What about the other planets?

Venus and Mars are the only two planets on which unmanned landings have been made, but all the planets except faraway Pluto have been visited by "flyby" missions. The U.S.-built *Voyager 1* and *Voyager 2* probes were both launched in 1977 on separate journeys to Saturn, Jupiter and the outer planets. The last planet to be visited was Neptune, in 1989, when *Voyager 2* flew past it on its way out of the solar system.

Living in space?

The recent adventures of astronauts and cosmonauts have been less dramatic. But they have been quietly perfecting the technology of space exploration, like transferring crew members from one vehicle to another, docking spacecraft with space stations, repairing satellites and moving around outside the spacecraft.

In the U.S., NASA (the National Aeronautics & Space Administration) introduced the first shuttle, *Columbia*, in 1981. This re-usable space plane, which can land like a glider on a special runway, has made it much cheaper to travel in space.

Meanwhile the Russian cosmonauts have concentrated on the problems of spending many months in space under weightless conditions. Some of their cosmonauts have spent over a year at a time in the *Mir* space station, which was launched in 1986.

Onwards and outwards

Now we know how to reach space and how to survive there, there is no doubt that humans will probe further and deeper before too long. NASA is planning for a human landing on Mars in the 2000s. On that day when Neil Armstrong stepped on to the moon, humanity started a whole new adventure in our exploration of the solar system and the universe.

Neil Armstrong (left), "Buzz" Aldrin (right) and Michael Collins, the three astronauts from the 1969 Apollo 11 *moon mission. Armstrong was the first human to set foot on the moon.*

Further Reading

Bernstein, Jeremy. *The Analytical Engine: Computers Past, Present and Future.* New York: Morrow, 1981.

Bornstein, Sandy, and Jerry Bornstein. *New Frontiers in Genetics.* Englewood Cliffs, NJ: Messner, 1984.

Brown, Julie, and Robert Brown. *Inventors.* London: Belitha Press, 1991.

Chaisson, Eric. *Cosmic Dawn.* Boston: Little, Brown, 1981.

Dolan, Edward F. *Inventors for Medicine.* New York: Crown, 1971.

Gies, Joseph, and Frances Gies. *The Ingenious Yankees.* New York: Crowell, 1976.

Gonick, Larry, and Art Huffman. *The Cartoon Guide to Physics.* New York: HarperCollins, 1990.

Graham, Ian. *Space Shuttles.* New York: Gloucester Press, 1989.

Jespersen, James, and Jane Fitz-Randolph. *RAMS, ROMS and Robots: The Inside Story of Computers.* New York: Atheneum, 1984.

Kennedy, George P. *The First Men in Space.* New York: Chelsea House, 1990.

Leakey, Richard. *The Making of Mankind.* New York: Dutton, 1981.

McGrave, Sharon Bertsch. *Nobel Prize Women in Science.* New York: Carol/Birch Lane Press, 1993.

McTavish, Douglas. *Famous Inventors.* Denver, CO: Wayland, 1993.

Moolan, Valerie. *The Road to Kitty Hawk.* New York: Time-Life Books, 1980.

Morgan, Nina. *Famous Scientists.* Denver, CO: Wayland, 1993.

Nardo, Don. *Charles Darwin.* New York: Chelsea House, 1993.

Pizer, Vernon. *The Irrepressible Automobile.* New York: Dodd, Mead, 1986.

Rathburn, Elizabeth. *Exploring Your Solar System.* Washington, D.C.: National Geographic Society, 1989.

Shore, Nancy. *Amelia Earhart.* New York: Chelsea House, 1987.

Spangenburg, Ray, and Diane Moser. *Opening the Space Frontier.* New York: Facts on File, 1989.

Sussman, Aaron. *The Amateur Photographer's Handbook.* New York: Harper and Row, 1973.

Tames, Richard. *The Wright Brothers.* New York: Franklin Watts, 1990.

Waitley, Douglas. *The Roads We Traveled: An Amusing History of the Automobile.* Englewood Cliffs, NJ: Messner, 1979.

Wescott, Lynanne, and Paula Degen. *Wind and Sand: The Story of the Wright Brothers at Kitty Hawk.* New York: Abrams, 1983.

Index

Abacus, 72
Acceleration, 26, 67–68
Advertisement, first, 18
Aerobatics, 59
Aging, relativity and, 67
Agnelli, Giovanni, 61
Aiken, Howard, 72
Aircraft, 55–59, 68
Air mail, 58
Alcock, John, 58
Aldrin, Buzz, 86, 90
Analytical Engine, Babbage's, 72
Antiseptics, 44–47
Apollo space program, 86–87, 90
Appendicitis, 45
Appert, Nicolas, 32–34
Archaeopteryx, 42
Arlandes, Marquis d', 59
Armillary sphere, 21
Armstrong, Neil, 86, 90
Artificial hip replacement, 45
Asepsis, 47
Astronauts, 82–83, 86–87
Astronomy, 20–22
  observing eclipses, 35
  photography and, 38
  satellites for, 83–84
Atom: splitting, 75
  structure, 75
Atom bomb, 54, 67, 68, 75–78

Babbage, Charles, 72–74
Bacon, Roger, 11
Bacteria, killing, 32
Baird, John Logie, 69–70
Baked beans, 34
Barnard, Dr. Christiaan, 45
Battering ram, 10
Beagle, HMS (ship), 39
Bed sores, 45
Bell, Alexander Graham, 48–51
Benz, Carl, 60
Berg, Madame, 59
Bible: first printed, 17, 18–19
  theory of evolution and, 40, 42, 43
"Big Bertha" (cannon), 11
Bonaparte, Napoleon, 32
Books, 15–19
  first with photographs, 36
Borden, Gail, 34
Boulton, Matthew, 28, 29
"Box camera," 35
Broadsides (song sheets), 19
Brown, Arthur Whitten, 58
Business Revolution, 30–31

Calculating machines, 72–74
Calculus, 26
Calvert, Crace, 45
Camera, 35–36
  television, 69–70
Camera obscura, 35
Cameron, Julia Margaret, 36
Cancer: radiation and, 77
  treatments, 47, 54
Canned food, 32–34

Cannon, 10–11
Carbolic acid, 45
Carbon-14, 77
Car industry, 60–64
Cathode-ray tube, 69, 70
Caxton, William, 19
Central heating, 30
"Chain reaction," 75, 78
Chapbooks, 19
Chernobyl, Ukraine, 77
Chromosomes, 79
Church: and books, 19, 22
  and Copernicus, 21
  and Galileo, 21, 22
  Inquisition, 22
  and printing, 18–19
  and theory of evolution, 42–43
Clock, pendulum, 22
Coal mining, 31
  water pump, *see* Newcomen engine
Coelacanth (fish), 43
Coin minting, 17
Collins, Michael, 86, 90
Colossus I (computer), 72
Comets, 21
  Moon and, 26
Computer graphics, 74
Computers, 72–74
  first voice-operated, 74
Concorde (aircraft), 59
Condensed milk, 34
Condenser, 28
Contact prints, 36
"Convenience food," 34
Copernicus, Nicolaus, 20–21
Copying machine, 30
Cosmonauts, 87
Cox, R.B., 74
Creation, date of, 42
Crick, Francis, 79–80
Crossbows, 14
Cuba missile crisis, 77
Cugnot, Joseph, 31
Curie, Irene, 54
Curie, Marie and Pierre, 52–54

Daguerre, Louis Jacques, 35, 36
Daguerreotypes, 36, 38
Da Vinci, Leonardo, 55
Darwin, Charles, 39–43
Darwin, Erasmus, 40
De Revolutionibus Orbium Coelestium
  (book by Copernicus), 21
Disabilities, evolution and, 43
Discourse on Light (book by Newton), 24
Diseases: evolution and, 43
  vaccines against, 47
DNA (Deoxyribonucleic acid), 79–80
Donkin, Bryan, 32, 34
Double helix, 80
Dressings, antiseptic, 45
Dunant, Henri, 11

Earhart, Amelia, 58
Early Bird I (satellite), 84
Eastman, George, 35–36
Edison, Thomas Alvar, 48
Eggestein, Heinrich, 18
Einstein, Albert, 26, 65–68
Electricity generation, nuclear, 78
Energy: nuclear (atomic), 67, 75, 78
  relativity and, 67
ENIAC (computer), 74
"Enigma" encoding machine, 72, 74
Environmental damage (pollution), 31
  evolution and, 43
Evolution, Theory of, 39–43
  ban on teaching, 42
Experiments: atomic, 75
  bad beer, 44–45
  evolution, 42
  gravity, 20, 21, 24–26, 68
  motion, 24
  optical, 24
  pendulum, 21, 22
  velocity of light, 65–67
Explorer I (spacecraft), 82

Face masks, surgical, 45, 47
Factories: canning, 32
  conditions, 31
  mechanized, 30, 31
  production lines, 60, 61, 63–64
Fermi, Enrico, 78
Film, photographic, 35–36
Finches, Galápagos Islands species,
  39–40
Fireworks, 11
Fluxions, *see* Calculus
"Flybys," 87–88, 90
Flyer (aircraft), 55
Flying boat, 58
Food: explorers' supplies, 34
  preservation, 32–34
Force, international unit of, 26
Ford, Henry, 60–64
Fossils, 42
  dating, 77
Fox Talbot, William Henry, 36
Franklin, Rosalind, 79–80
Fust, Johann, 17

Gagarin, Yuri, 82, 87
Galápagos Islands survey, 39, 40
Galileo, 20–22
Genes, 79
Genetic engineering, 80
"Genetic finger-printing," 80
Genetics: evolution and, 43
  radiation and, 77
Germs: discovery of, 44–45
  killing, 45–47
Glenn, John, 82
Gliders, 56
Gloves, surgical, 45, 47
Gould, John, 39
Governor (engine regulator), 30
Gravity, 20, 21, 24–26
  Einstein's theory of, 67–68
  and straight lines, 68
Gray, Elisha, 48
Great Plague, 24
Gunpowder, 10–14
Gutenberg, Johannes, 15, 17

Hahn, Otto, 75
Hainy, John, 45
Halftone process, 35
Hall, John, 32, 34
Halley, Edmund, 26
Handguns, 11–12
Harvard Mark I (computer), 74
Heart transplants, 45, 47
Heavier-than-air machine, 55
Heinz, 34
Helicopter, first, 59
Henri Grâce à Dieu (flagship), 14
Heredity: DNA and, 79
  evolution and, 43
Hipparchus, 20
Hiroshima, Japan, 75–77
Hot-air balloon, 55–56, 59
Hubble Telescope, 84
Hydrogen bomb, 78

IBM Mk7 (electronic calculator), 72
Industrial Revolution, 28, 30–31
Infections, preventing, 44–47
"Instant camera", *see* Polaroid camera
Integrated circuits, 74
Intelsat (satellite), 83
International Red Cross, 11
International Ultra Violet Explorer (space
  telescope), 83

Jet engines, 59
Johannsen, Wilhelm, 79
Johnson, Amy, 58
Joliot-Curie, Frederic, 54

Key-opening food can, 34
Kites, 55–56
Kleitsch, Karl, 35

Labrosse-Garnerin, Jeanne, 59
Laika (dog), 82
Lamarck, Chevalier de, 40
Land, Edwin, 36
Landsat 4 (satellite), 83
Lenses: camera, 35
  telescope, 22
  television camera, 69
Light-pen, 74
Light velocity (speed), 65
  measuring accurately, 68
Lindbergh, Charles, 58
Lister, Dr. Joseph, 44–47
Longbows, 14
Luna space probes, 86
Lunik II (spacecraft), 86

Machinery, steam-driven, 30
Magellan (satellite), 88
Manhattan Project, 75
Mariner 2 (space probe), 87
Martins, B., 74
Mass, relativity and, 65–67
Mass production: cars, 62–64, 74
Master gunner, 10, 12
Mechanics (physics), 22
Megatherium (giant sloth), 42
Meitner, Lise, 75
Mendel, Gregor, 79, 80
Merry Oldsmobile, 60, 64
Micro-organisms, *see* Germs
Microscope, Darwin's, 43
Milky Way, 21
Miner's Friend, 30
Mir (space station), 90
Model T Ford, 61–64